Large Time Behavior
of Solutions for General
Quasilinear Hyperbolic-Parabolic
Systems of Conservation Laws

MEMOIRS
of the
American Mathematical Society

Number 599

Large Time Behavior
of Solutions for General
Quasilinear Hyperbolic-Parabolic
Systems of Conservation Laws

Tai-Ping Liu
Yanni Zeng

January 1997 • Volume 125 • Number 599 (end of volume) • ISSN 0065-9266

American Mathematical Society
Providence, Rhode Island

1991 *Mathematics Subject Classification.*
Primary 35K55, 76N10; Secondary 35B40, 35A08, 35L65, 76W05.

Library of Congress Cataloging-in-Publication Data

Liu, Tai-Ping, 1945–
 Large time behavior of solutions for general quasilinear hyperbolic-parabolic systems of conservation laws / Tai-Ping Liu, Yanni Zeng.
  p. cm.—(Memoirs of the American Mathematical Society, ISSN 0065-9266 ; no. 599)
 "January 1997, volume 125, number 599 (end of volume)."
 Includes bibliographical references.
 ISBN 0-8218-0545-2 (alk. paper)
  1. Shock waves. 2. Conservation laws (Mathematics) 3. Differential equations, Parabolic—Numerical solutions. 4. Differential equations, Hyperbolic—Numerical solutions. I. Zeng, Yanni, 1955– . II. Title. III. Series.
QA3.A57   no. 599
[QA927]
510 s—dc20
[531′.1133]
                                                                      96-44759
                                                                      CIP

Memoirs of the American Mathematical Society

This journal is devoted entirely to research in pure and applied mathematics.

Subscription information. The 1997 subscription begins with number 595 and consists of six mailings, each containing one or more numbers. Subscription prices for 1997 are $414 list, $331 institutional member. A late charge of 10% of the subscription price will be imposed on orders received from nonmembers after January 1 of the subscription year. Subscribers outside the United States and India must pay a postage surcharge of $30; subscribers in India must pay a postage surcharge of $43. Expedited delivery to destinations in North America $35; elsewhere $110. Each number may be ordered separately; *please specify number* when ordering an individual number. For prices and titles of recently released numbers, see the New Publications sections of the *Notices of the American Mathematical Society.*

Back number information. For back issues see the *AMS Catalog of Publications.*

Subscriptions and orders should be addressed to the American Mathematical Society, P. O. Box 5904, Boston, MA 02206-5904. *All orders must be accompanied by payment.* Other correspondence should be addressed to Box 6248, Providence, RI 02940-6248.

Memoirs of the American Mathematical Society is published bimonthly (each volume consisting usually of more than one number) by the American Mathematical Society at 201 Charles Street, Providence, RI 02904-2294. Periodicals postage paid at Providence, RI. Postmaster: Send address changes to Memoirs, American Mathematical Society, P. O. Box 6248, Providence, RI 02940-6248.

CONTENTS

ABSTRACT

We are interested in the time-asymptotic behavior of solutions to viscous conservation laws. Through the pointwise estimates for the Green's function of the linearized system and the analysis of coupling of nonlinear diffusion waves, we obtain explicit expressions of the time-asymptotic behavior of the solutions. This yields optimal estimates in the integral norms. For most physical models, the viscosity matrix is not positive definite and the system is hyperbolic-parabolic, and not uniformly parabolic. This implies that the Green's function may contain Dirac δ-functions. When the corresponding inviscid system is non-strictly hyperbolic, the time-asymptotic state contains generalized Burgers solutions. These are illustrated by applying our general theory to the compressible Navier-Stokes equations and the equations of magnetohydrodynamics.

1991 *Mathematics Subject Classification*: Primary 35K55, 76N10; Secondary 35B40, 35A08, 35L65, 76W05.

Key words and phrases: quasilinear hyperbolic-parabolic systems, large time behavior, Green's function, nonlinear and linear diffusion waves, conservation laws, compressible Navier-Stokes equations, magneto-hydrodynamics.

The research of the first author was partially supported by Army Research Grant DAAH04-94-G-0045 and NSF Grant DMS-9216275-001.

The research of the second author was supported by the Office of Naval Research under contract # N00014-92-J-1481. Additional support was also provided by the Army Research Office under contract # DAAH04-93-G-0125 and by the National Science Foundation under grant # DMS-9307928.

This monograph was prepared in $\mathcal{A}_{\mathcal{M}}\mathcal{S}$-TEX.

1. Introduction

Consider a system of viscous conservation laws

$$(1.1) \qquad u_t + f(u)_x = (B(u)u_x)_x,$$

where $u = u(x,t)$ is the density of physical quantities, $f(u)$ is the flux and $B(u)$ is the viscosity matrix. Many physical models are of this form, e.g. the compressible Navier-Stokes equations, the equations of magnetohydrodynamics and nonlinear viscoelasticity models. We are interested in the explicit, pointwise large time behavior of the solution of the initial value problem of the system. The solution approaches, time-asymptotically, a linear superposition of linear heat kernels, generalized Burgers solutions, and diffusion waves of algebraic types.

Several features of the system contribute to the rich qualitative behavior of the solution. The first basic feature is the nonlinearity of the flux function $f(u)$. Physical models are symmetrizable, which implies that the inviscid system

$$(1.2) \qquad u_t + f(u)_x = 0$$

is completely hyperbolic with real characteristics $\lambda_1(u) \le \lambda_2(u) \le \cdots \le \lambda_n(u)$:

$$(1.3) \qquad \begin{aligned} f'(u)r_i(u) &= \lambda_i(u)r_i(u), & l_i(u)f'(u) &= \lambda_i(u)l_i(u), \\ l_i(u)r_j(u) &= \delta_{ij}, & i,j &= 1,2,\dots,n. \end{aligned}$$

The first important nonlinearity is the behavior of the characteristic speed $\lambda_i(u)$ in the wave direction $r_i(u)$. An i-field is genuinely nonlinear (g.nl.) if $\lambda_i(u)$ is strictly monotone, linearly degenerate (l.dg.) if it is stationary, [La],

$$\text{(g.nl.)} \qquad \nabla\lambda_i(u) \cdot r_i(u) \neq 0,$$

$$\text{(l.dg.)} \qquad \nabla\lambda_i(u) \cdot r_i(u) = 0.$$

Waves pertaining to a (g.nl.) field either compress or expand. Compression waves lead to shock waves. A (l.dg.) field gives rise to linear hyperbolic waves, the contact discontinuities.

Received by the editor August 29, 1994; and in revised form October 30, 1995.

1

For the viscous system (1.1), (g.nl.) fields yield nonlinear diffusion waves, solutions of Burgers' equation; while (l.dg.) fields yield the linear heat kernels as the time-asymptotic state.

The second nonlinearity is the coupling of waves of different characteristic families. This is measured by

$$l_i f''(r_j, r_j), \qquad i \neq j.$$

For physical models, these quantities are nonzero. This implies that the linear and nonlinear diffusion waves interact nonlinearly and produce diffusion waves which decay algebraically in x and t.

Another nonlinearity is the dependence of the multiplicity of the characteristic values $\lambda_i(u)$, $i = 1, 2, \ldots, n$, on the state u. The compressible Euler equations are always strictly hyperbolic. But many other models, including the equations of magnetohydrodynamics, full elasticity system and multiphase flows are nonstrictly hyperbolic. For instance, Alfvén, slow and fast acoustic speeds may equal for the equations of magnetohydrodynamics, see [LLP]. When the inviscid model is nonstrictly hyperbolic, system (1.1) possesses time-asymptotic states of solutions of generalized Burgers equations. These are systems with single characteristic speed and quadratic nonlinear flux.

The second basic feature of the viscous conservation laws (1.1) is that physical viscosity matrix $B(u)$ is not diagonalizable with the inviscid system, and not even positive definite. This is dictated by physics. For instance, the continuity equation is not dissipative. Physical models, nevertheless, are dissipative when the relevant dissipation parameters are present. The general mathematical hypotheses are to assume that there exists a dissipative symmetrizer and that no characteristic direction of the flux is in the null space of the viscosity matrix. With such physical assumptions, the hyperbolic nature of system (1.1) is reflected in the propagation of the discontinuities in the initial data into the solution at all time. This is reflected in the present study as Dirac δ-functions contained in the Green's function for the linearized system of (1.1). We have explicit expressions for the characteristic information on the δ-functions and the rate of exponential decay. This non-smoothing property complicates our analysis on large time behavior. That the viscosity matrix is not diagonalizable with the Jacobi matrix of the flux means that there is a coupling of the heat kernels in the Green's function. We show that the Green's function equals that for a diagonalizable system plus faster decaying terms.

The present paper is concerned with the large time behavior of solutions of (1.1) that are perturbations of a constant state. Without loss of generality, we take the constant state to be the zero state,

$$(1.4) \qquad u(x,0) = u_0(x), \qquad u_0(\pm\infty) = 0.$$

Our approach contains two new major elements, the accurate pointwise estimate on the Green's function of the linearized equations

$$(1.5) \qquad u_t + f'(0)u_x = B(0)u_{xx},$$

and the analysis on the nonlinear coupling of diffusion waves. We now briefly describe the setup. Since the system is dissipative, the solution is expected to decay to the zero state in the sup norm. Expand the solution around the zero state, we have from (1.1) that

$$u_{it} + \left[\lambda_i(0)u_i + \sum_{j,k=1}^{n} \frac{1}{2}l_i(0)f''(0)(u_j r_j(0), u_k r_k(0)) \right]_x$$

$$(1.6) \qquad = \sum_{j=1}^{n} [l_i(0)B(0)(u_j r_j(0))_x]_x + [O(1)(|u|^3 + |u||u_x|)]_x, \qquad i=1,\ldots,n,$$

$$u(x,t) = \sum_{i=1}^{n} u_i(x,t)r_i(0).$$

The first step is to approximate this by a simpler system, which yields explicit solutions. For strictly hyperbolic case, $\lambda_1(0) < \lambda_2(0) < \cdots < \lambda_n(0)$, we ignore all the coupling terms of different families and the higher order terms beyond the quadratic terms to obtain

$$(1.7) \quad \theta_{it} + \lambda_i(0)\theta_{ix} + [\frac{1}{2}l_i(0)f''(0)(r_i(0),r_i(0))\theta_i^2]_x = l_i(0)B(0)r_i(0)\theta_{ixx}, \quad i=1,\ldots,n.$$

This is the linear heat equation or Burgers equation. For nonstrictly hyperbolic case, the same procedure yields a system of equations similar to (1.7). These equations are called generalized Burgers equations, where all the modes corresponding to the same characteristic speed at $u = 0$ are coupled. In any case, the approximate equations, (1.7) or generalized Burgers equations, possess self-similar solutions. These are the linear and nonlinear heat kernels:

$$(1.8) \qquad \theta_i(x,t) = t^{-\frac{1}{2}}\chi_i(\frac{x - \lambda_i(0)t}{\sqrt{t}})$$

for some functions χ_i. These solutions represent the time-asymptotic state for the approximate equations. They are also the main term in the time-asymptotic state for the solution
of (1.1). Equation (1.8) represents a one-parameter family of solutions with the parameter

$$c_i = \int_{-\infty}^{\infty} \theta_i(x,t)\,dx.$$

Choose c_i so that

$$\int_{-\infty}^{\infty} v(x,t)\,dx = 0, \qquad v(x,t) \equiv u(x,t) - \sum_{i=1}^{n} \theta_i(x,t) r_i(0).$$

For simplicity, we use the strictly hyperbolic case to illustrate what we need to estimate
v and the structure of the paper. Let $G(x,t)$ be the Green's function associated with the
linearization of (1.6), and $G^*(x,t)$ be the Green's function associated with the linearization
of (1.7). Clearly G^* is a diagonal matrix whose ith diagonal element G_{ii}^* is a heat kernel
along the $\lambda_i(0)$ direction. We have from Duhamel's principle that

$$
\begin{aligned}
v_i(x,t) = & \int_{-\infty}^{\infty} G_{ii}^*(x-y,t)v_i(y,0)\,dy + \int_{-\infty}^{\infty} (G-G^*)_i(x-y,t)\check{u}(y,0)\,dy \\
& + \int_0^t \int_{-\infty}^{\infty} G_{ii}^*(x-y,t-\tau)\Bigg[\sum_{j\neq i} c_{ij}\theta_j^2 + \sum_{j=1}^{n} 2c_{ij}\theta_j v_j + O(1)|v|^2 \\
& \hspace{4cm} + \sum_{j\neq k} O(1)\theta_j(\theta_k+v_k)\Bigg]_y (y,\tau)\,dy d\tau \\
& + \int_0^t \int_{-\infty}^{\infty} (G-G^*)_i(x-y,t-\tau)\big[O(1)(|\theta|^2+|v|^2)\big]_y(y,\tau)\,dy d\tau \\
& + \int_0^t \int_{-\infty}^{\infty} G_i(x-y,t-\tau)\big[O(1)(|\theta|^3+|v|^3 \\
& \hspace{3cm} + (|\theta|+|v|)(|\theta_y|+|v_y|))\big]_y(y,\tau)\,dy d\tau, \\
& i = 1,\dots,n, \\
v(x,t) \equiv & \sum_{i=1}^{n} v_i(x,t) r_i(0),
\end{aligned}
$$

(1.9)

where $(G-G^*)_i$ and G_i denote the ith row of $G-G^*$ and G respectively, c_{ij}, i, $j = 1$,
..., n, are constants, and

$$\check{u} = (u_1,\dots,u_n)^t, \qquad \theta = (\theta_1,\dots,\theta_n)^t.$$

A more precise expression of v in the general case (not necessary to be strictly hyperbolic) is derived in Section 8, while the exact formulation of our main theorems about the estimates of v is contained in Section 2.

To make use of (1.9) we need to have accurate estimates of the Green's function G and the convolutions of the heat kernel G_{ii}^* with the sources $\sum_{j\neq i} \theta_j^2$ and the nonlinear term v^2. The latter is the study of nonlinear coupling of diffusion waves. The diffusion waves are either of the linear or nonlinear heat kernel types (1.8) or of algebraic types

(1.10)
$$\psi_{3/2}(x,t;\lambda_i(0)) = [(x - \lambda_i(0)(t+1))^2 + t + 1]^{-3/4},$$
$$\tilde{\psi}(x,t;\lambda_i(0)) = [|x - \lambda_i(0)(t+1)|^3 + (t+1)^2]^{-1/2},$$
$$(\phi_1\psi_{3/4})(x,t;\lambda_i(0)) = (|x - \lambda_i(0)(t+1)| + t + 1)^{-1/2}$$
$$\times [(x - \lambda_i(0)(t+1))^2 + t + 1]^{-3/8}.$$

In Section 3 we study the dissipation of these waves. Our analysis is motivated by and greatly generalize that of [Liu], where diffusion waves of algebraic types were identified.

For a diagonalizable system, the Green's function consists of heat kernels. The time-asymptotic analysis for this simplified case is carried out in Section 4.

Physical viscosity matrix is non-positive definite, and, as a result, the Green's function contains Dirac δ-functions. Moreover, the system is nondiagonalizable and so the Green's function differs from that for a diagonalizable system also by a term which decays at a faster rate. These phenomena were found in [Ze] for the isentropic compressible Navier-Stokes equations. In Section 5, we study the same problem but using a simpler method. This new method can be extended to general systems; see Section 6.

In our study of dissipation nature of the linearization of (1.6), or equivalently the linear system (1.5), through the pointwise estimates of its Green's function, we make uses of Kato's perturbation theory, [Kt], to analyze the singularities of the spectrum of the matrix in Fourier variables. In consistence with physical models, we assume that (1.1) is symmetrizable. That is, it is endowed with an entropy function. This removes the possible singularity of the spectrum at the origin or at the infinity. We then show that the Green's function G for (1.5) contains G^*, the heat kernels corresponding to the associated diagonalizable system. Again, motivated by physical consideration, we assume that the null space of the viscosity matrix B does not contain the inviscid characteristic directions.

This guarantees the dissipativeness of the system. We show that the remainder $G - G^*$ decays at a faster rate than that for G^*. The remainder $G - G^*$ also contains exponentially decaying Dirac δ-functions due to the hyperbolic nature of our system. In Section 6, we also apply this general theory to the compressible Navier-Stokes equations and the equations of magnetohydrodynamics. We calculate out explicitly the heat kernels as well as the Dirac δ-functions contained in G for these equations.

For the isentropic compressible Navier-Stokes equations, it has been shown that discontinuities in the initial data propagate into the flow along the particle path, [H]. This corresponds to the δ-function in the Green's function. For the full compressible Navier-Stokes equations, we also find that there is a δ-function along the particle path when both viscosity and heat conductivity coefficients are positive. When the viscosity coefficient is zero, this δ-function splits into two δ-functions about, but not on, the particle path. For the equations of magnetohydrodynamics, there may exist three δ-functions.

Section 7 concerns the energy estimate due to Kawashima [Ka]. Since the presentation is short, we present it here for completeness. For the estimate we need the additional hypothesis of invariance of the null space of the viscosity matrix with respect to the symmetrizer; see Assumption 2.2 of Section 2. The energy estimate is needed to close the stability analysis for the general system, which is carried out in Section 8.

Finally, we verify the aforementioned dissipativeness hypotheses for the compressible Navier-Stokes equations and the equations of magnetohydrodynamics in Section 9.

L^p optimal convergence rate of v in (1.9) has been obtained by Zeng, [Z], which generalized her previous work on p-system, [Ze]. Here our new pointwise approach yields explicit information on the propagation of diffusion waves. The analysis is effective in studying the coupling of these waves. Many physical models are only partially dissipative. This is so, for instance, for the compressible Navier-Stokes equations with positive viscosity and zero heat conductivity, and also for the compressible Euler equations with thermal nonequilibrium. For such a system, there are diffusive and non-dissipative waves. We believe that our analysis can be generalized to study the behavior of these waves. This is, however, left to the future.

2. VISCOUS CONSERVATION LAWS

Consider the Cauchy problem of a general system of viscous conservation laws:

(2.1) $u_t + f(u)_x = (B(u)u_x)_x,$ $-\infty < x < \infty,$ $t > 0,$

(2.2) $u(x,0) = u_0(x),$ $-\infty < x < \infty,$

where $u = u(x,t)$ is an n-vector, $f(u)$ is a smooth n-vector valued function, $B(u)$ a smooth $n \times n$ matrix, and $u_0(x)$ a given n-vector valued function. We assume that u_0 is a small perturbation of a constant state. Without loss of generality, we assume that the constant state is zero. To assure that system (2.1) is dissipative we make the following three basic assumptions:

Assumption 2.1. System (2.1) has a strictly convex entropy U ([Sm], [K]).

Assumption 2.1 means that there exist a pair of smooth functions $U(u)$ and $F(u)$ (called entropy pair), such that U is strictly convex, $(\nabla U)f' = \nabla F$, and $(\nabla^2 U)B$ is symmetric and semi-positive definite, where f' is the Jacobi matrix of f, and $\nabla^2 U$ is the Hessian of U. We set $A_0(u) = \nabla^2 U(u)$. Clearly A_0 is symmetric and positive definite.

Assumption 2.2. For small u, $B(u) \neq 0$. There exists a smooth one-to-one mapping $u = f_0(\tilde{u})$, $f_0(0) = 0$, such that the null space $\mathcal{N}$ of $\tilde{B}(\tilde{u}) \equiv B(f_0(\tilde{u}))f_0'(\tilde{u})$ is independent of $\tilde{u}$. Moreover, $\mathcal{N}^\perp$ is invariant under $f_0'(\tilde{u})^t A_0(f_0(\tilde{u}))$, and $\tilde{B}(\tilde{u})$ maps $\mathbb{R}^n$ into $\mathcal{N}^\perp$, where $\mathcal{N}^\perp$ is the orthogonal complementement of $\mathcal{N}$.

Assumption 2.3. Any eigenvector of $f'(0)$ is not in the null space of $B(0)$.

We notice that several interesting physical models, e.g. the Navier-Stokes equations and the equations of magnetohydrodynamics, satisfy Assumptions 2.1–2.3 (c.f. Section 9). We also noticc that if $B(u)$ is nonsingular, Assumptions 2.2 and 2.3 are automatically satisfied.

Under Assumption 2.1, it follows that $A_0 f'$ is symmetric [FL]. Since A_0 is symmetric and positive definite, there exists a symmetric positive definite matrix $A_0^{\frac{1}{2}}$ such that $A_0 = (A_0^{\frac{1}{2}})^2$. Thus $A_0^{\frac{1}{2}} f'(A_0^{\frac{1}{2}})^{-1}$ is symmetric. Consequently, all the eigenvalues of $A_0^{\frac{1}{2}} f'(A_0^{\frac{1}{2}})^{-1}$ are real, and $A_0^{\frac{1}{2}} f'(A_0^{\frac{1}{2}})^{-1}$ has a complete set of eigenvectors. Notice that if $A_0^{\frac{1}{2}} f'(A_0^{\frac{1}{2}})^{-1}$ has an eigenvalue λ and a corresponding eigenvector r, $A_0^{\frac{1}{2}} f'(A_0^{\frac{1}{2}})^{-1}r = \lambda r,$

then $f'(A_0^{\frac{1}{2}})^{-1}r = \lambda(A_0^{\frac{1}{2}})^{-1}r$, i.e., λ is an eigenvalue of f' with $(A_0^{\frac{1}{2}})^{-1}r$ as a corresponding eigenvector. For this reason f' has real eigenvalues $\bar{\lambda}_1(u) \leq \bar{\lambda}_2(u) \leq \cdots \leq \bar{\lambda}_n(u)$ and a complete set of eigenvectors:

$$(2.3) \qquad \begin{aligned} f'(u)\bar{r}_i(u) &= \bar{\lambda}_i(u)\bar{r}_i(u), \\ \bar{l}_i(u)f'(u) &= \bar{\lambda}_i(u)\bar{l}_i(u), \\ \bar{l}_i(u)\bar{r}_j(u) &= \delta_{ij}, \qquad i,j,= 1,2,\ldots,n. \end{aligned}$$

In other words, the inviscid system

$$(2.4) \qquad\qquad u_t + f(u)_x = 0$$

corresponding to (2.1) is completely hyperbolic.

We are interested in the perturbation of the zero state. Hence the eigenvalues and eigenvectors of $f'(0)$ play an important role. We use special notations to denote them. Let $\lambda_1 < \lambda_2 < \cdots < \lambda_s$ be the distinct eigenvalues of $f'(0)$ with multiplicity $m_1, m_2, \ldots, m_s$, $m_1 + m_2 + \cdots + m_s = n$. Let the left and right eigenvectors associated with λ_i be l_{ij} and $r_{ij}, j = 1, \ldots, m_i$:

$$(2.5) \qquad \begin{aligned} f'(0)r_{ij} &= \lambda_i r_{ij}, \\ l_{ij}f'(0) &= \lambda_i l_{ij}, \\ l_{ij}r_{i'j'} &= \delta_{ii'}\delta_{jj'}, \\ i,i' &= 1,\ldots,s, \quad j = 1,\ldots,m_i, \quad j' = 1,\ldots,m_{i'}. \end{aligned}$$

Introduce the notations

$$(2.6) \qquad l_i = \begin{pmatrix} l_{i1} \\ \vdots \\ l_{im_i} \end{pmatrix}, \qquad r_i = (r_{i1},\ldots,r_{im_i}), \qquad i = 1,\ldots,s.$$

Clearly,

$$(2.7) \qquad l_i r_i = I_{m_i \times m_i}, \qquad l_i r_j = 0_{m_i \times m_j}, \qquad i \neq j.$$

Decompose the solution u to (2.1), (2.2) in the right eigenvector directions,

$$(2.8) \qquad u(x,t) = \sum_{i=1}^{s}\sum_{j=1}^{m_i} r_{ij}u_{ij}(x,t) = \sum_{i=1}^{s} r_i u_i(x,t),$$

where

$$(2.9) \qquad u_i(x,t) = (u_{i1}, \ldots, u_{im_i})^t(x,t) = l_i u(x,t), \qquad i = 1, \ldots, s.$$

Our main result says that $u_i = \theta_i + v_i$, where θ_i is a self-similar solution to the Burgers equation, the heat equation , or the generalized Burgers equation, while the remainder v_i decays more rapidly in t.

To give a heuristic explanation of the construction of the ansatz θ_i, we first consider the equation satisfied by u_i. With (2.9) and (2.8), multiply (2.1) by l_i, and take Taylor expansion about zero to obtain

$$(2.10) \quad u_{it} + \lambda_i u_{ix} + \frac{1}{2} l_i f''(0) (\sum_j r_j u_j, \sum_j r_j u_j)_x$$

$$= l_i B(0) \sum_j r_j u_{jxx} + (O(1)(|u||u_x| + |u|^3))_x, \qquad i = 1, \ldots, s.$$

Each of the above equations is associated with an eigenvalue λ_i. Neglecting the effects of the other families and the higher order terms, we have the following approximate equations:

$$(2.11) \qquad \theta_{it} + \lambda_i \theta_{ix} + \frac{1}{2} l_i f''(0)(r_i \theta_i, r_i \theta_i)_x = l_i B(0) r_i \theta_{ixx}, \qquad i = 1, \ldots, s,$$

where θ_i is an m_i-vector. The ansatz θ_i of u_i is defined as the self-similar solution to (2.11) carrying the same mass as u_i:

$$(2.12) \qquad \int_{-\infty}^{\infty} \theta_i(x,t)dx = \int_{-\infty}^{\infty} l_i u_0(x)dx \equiv \delta_i.$$

We call θ_i, $i = 1, \ldots, s$, the diffusion waves of (2.1), (2.2). The following lemma shows that (2.11) is uniformly parabolic, though the full system (2.1) may be only hyperbolic-parabolic.

Lemma 2.1. [SK] *Under Assumptions 2.1 and 2.3, each $l_i B(0) r_i$ is similar to an $m_i \times m_i$ symmetric positive definite matrix, $i = 1, \ldots, s$.*

Proof. Let l'_{ij} and r'_{ij}, $i = 1, \ldots, s$, $j = 1, \ldots, m_i$ be other sets of left and right eigenvectors, respectively, satisfying (2.5). Using the notations in (2.6) it is clear that

$$l'_i = c_i l_i, \qquad r'_i = r_i d_i,$$

where c_i and d_i are $m_i \times m_i$ matrices. Since both pairs $\{l_i, r_i\}$ and $\{l'_i, r'_i\}$ satisfy (2.7), we have $d_i = c_i^{-1}$. Thus

$$l'_i B(0) r'_i = c_i l_i B(0) r_i c_i^{-1},$$

i.e., $l'_i B(0) r'_i$ is similar to $l_i B(0) r_i$. The lemma is proved if we can construct l_{ij} and r_{ij}, $i = 1, \ldots, s$, $j = 1, \ldots, m_i$, such that each $l_i B(0) r_i$ is symmetric and positive definite.

Under Assumption 2.1, both $A_0 B$ and $A_0 f'$ are symmetric, and $A_0 B$ is semi-positive definite, where $A_0 = \nabla^2 U$. To simplify notations for the moment we set $A_0 \equiv A_0(0)$, $B \equiv B(0)$, etc. As we have pointed out before, $\{r_{ij}\}$ are right eigenvectors of f' if and only if $\{A_0^{\frac{1}{2}} r_{ij}\}$ are right eigenvectors of $A_0^{\frac{1}{2}} f' (A_0^{\frac{1}{2}})^{-1}$, where $A_0^{\frac{1}{2}}$ is a symmetric positive definite matrix such that $A_0 = (A_0^{\frac{1}{2}})^2$. Similarly, $\{l_{ij}\}$ are left eigenvectors of f' if and only if $\{l_{ij}(A_0^{\frac{1}{2}})^{-1}\}$ are left eigenvectors of $A_0^{\frac{1}{2}} f'(A_0^{\frac{1}{2}})^{-1}$. We can choose $\{r_{ij}\}$ such that $\{A_0^{\frac{1}{2}} r_{ij}\}$ is an orthonormal set since $A_0^{\frac{1}{2}} f'(A_0^{\frac{1}{2}})^{-1}$ is symmetric. The left eigenvectors of $A_0^{\frac{1}{2}} f'(A_0^{\frac{1}{2}})^{-1}$ are then chosen as

$$l_{ij}(A_0^{\frac{1}{2}})^{-1} = (A_0^{\frac{1}{2}} r_{ij})^t,$$

or

$$l_{ij} = r_{ij}^t A_0,$$

$i = 1, \ldots, s$, $j = 1, \ldots, m_i$. Clearly $\{l_{ij}\}$ and $\{r_{ij}\}$ satisfy (2.5). Let $R = (r_1, \ldots, r_s)$. then each $l_i B r_i$ is an $m_i \times m_i$ principal submatrix of $R^t A_0 B R$. Since $R^t A_0 B R$ is symmetric and semi-positive definite, so is $l_i B r_i$. If $l_i B r_i$ were not positive definite, there would be a $w \in \mathbb{R}^{m_i}$, $w \neq 0$, such that

$$w^t l_i B r_i w = (r_i w)^t A_0 B r_i w = 0.$$

Thus

$$A_0 B r_i w = 0,$$

$$B(r_i w) = 0.$$

Here $r_i w \neq 0$ would be an eigenvector of f' corresponding to λ_i. This contradicts Assumption 2.3.   □

Consider the special case that λ_i is simple, $m_i = 1$. Then θ_i is a scalar and (2.11) becomes

(2.13) $$\theta_{it} + \lambda_i \theta_{ix} + c_{ii}(\theta_i^2)_x = \mu_i \theta_{ixx},$$

where

(2.14) $$c_{ii} = \frac{1}{2} l_i f''(0)(r_i, r_i), \qquad \mu_i = l_i B(0) r_i.$$

By Lemma 2.1, $\mu_i > 0$. Equation (2.13) is the heat equation if $c_{ii} = 0$. The corresponding diffusion wave is then the heat kernel:

(2.15) $$\theta_i(x, t) = \frac{\delta_i}{\sqrt{4\pi\mu_i(t+1)}} e^{-\frac{(x-\lambda_i(t+1))^2}{4\mu_i(t+1)}},$$

with δ_i given by (2.12). If $c_{ii} \neq 0$, equation (2.13) is the Burgers equation, which can be solved by Hopf-Cole transformation [Ho], [Co],

$$\theta_i = -\frac{\mu_i}{c_{ii}} (\ln \tilde{\theta}_i)_x,$$

$$\tilde{\theta}_{it} + \lambda_i \tilde{\theta}_{ix} = \mu_i \tilde{\theta}_{ixx}.$$

The diffusion wave is

(2.16) $$\theta_i(x, t) = \frac{\sqrt{\mu_i}}{2c_{ii}} (t+1)^{-\frac{1}{2}} (e^{\frac{\delta_i c_{ii}}{\mu_i}} - 1) e^{-\frac{(x-\lambda_i(t+1))^2}{4\mu_i(t+1)}}$$
$$\cdot \left[\sqrt{\pi} + (e^{\frac{\delta_i c_{ii}}{\mu_i}} - 1) \int_{\frac{x-\lambda_i(t+1)}{\sqrt{4\mu_i(t+1)}}}^{\infty} e^{-y^2} dy \right]^{-1}.$$

In (2.15) and (2.16) we have avoided the singularity of the heat kernel at $t = 0$ by replacing t by $t+1$. Notice from differentiating (2.3) that $l_i f''(0)(r_i, r_i) = \nabla \lambda_i(0) r_i$. Thus pertaining to a linearly degenerate field we have a linear diffusion wave (2.15), and to a genuinely nonlinear field we have a nonlinear diffusion wave (2.16), [La].

When λ_i is not simple, $m_i > 1$, system (2.11) is called the generalized Burgers equations [Ch]. The self-similar solution of (2.11), (2.12) is quite similar to the nonlinear diffusion wave (2.16). Precisely, we have the following lemma due to Chern.

Lemma 2.2. [Ch] *Under Assumptions 2.1 and 2.3, for small δ_i, (2.11), (2.12) has a unique self-similar solution of the form $\frac{1}{\sqrt{t}}\chi_i(\frac{x-\lambda_i t}{\sqrt{t}})$, with $\lim_{y\to-\infty} y\chi_i(y) = \chi'_i(-\infty) = 0$. Furthermore, χ_i has the property*

$$(2.17) \qquad \chi_i(y) = e^{-\frac{y^2}{4\mu_i}}\tilde{\chi}_i(y),$$

where μ_i is the maximum eigenvalue of $l_i B(0) r_i$, and $\tilde{\chi}_i$ and all its derivatives are uniformly bounded by $C|\delta_i|$, $C > 0$ is a constant.

The proof of Lemma 2.2 will be given in the proof of Lemma 3.1.

Again to avoid the singularity at $t = 0$, the diffusion wave pertaining to the i-family is then defined as

$$(2.18) \qquad \theta_i(x,t) = \frac{1}{\sqrt{t+1}}\chi_i\left(\frac{x-\lambda_i(t+1)}{\sqrt{t+1}}\right),$$

where $\frac{1}{\sqrt{t}}\chi_i(\frac{x-\lambda_i t}{\sqrt{t}})$ is the self-similar solution in Lemma 2.2.

Introduce the following notations,

$$\phi_\alpha(x,t;\lambda) = (|x-\lambda(t+1)|+t+1)^{-\alpha/2},$$

$$(2.19) \qquad \psi_\alpha(x,t;\lambda) = [(x-\lambda(t+1))^2 + t + 1]^{-\alpha/2},$$

$$\tilde{\psi}(x,t;\lambda) = [|x-\lambda(t+1)|^3 + (t+1)^2]^{-1/2}.$$

We call them generalized diffusion waves.

The main result of this paper is the following pointwise estimate of the solution to (2.1), (2.2).

Theorem 2.3. *Suppose that system (2.1) satisfies Assumptions 2.1–2.3, and that the initial data satisfy the following smoothness and smallness assumptions:*

$$(2.20) \qquad u_0 \in H^8(\mathbb{R}),$$

$$(2.21) \qquad v_-(x) \equiv \int_{-\infty}^x u_0(y)dy \in L^1(\mathbb{R}^-), \quad v_+(x) \equiv \int_x^\infty u_0(y)dy \in L^1(\mathbb{R}^+),$$

$$(2.22) \qquad u_0(x) = O(1)(1+|x|)^{-5/4}, \qquad u'_0(x) = O(1)(1+|x|)^{-5/4},$$

$$(2.23) \qquad v_\pm(x) = O(1)(1+|x|)^{-1},$$

$$
(2.24) \qquad
\begin{aligned}
&\|u_0\|_{H^8} + \|v_-\|_{L^1(-\infty,0)} + \|v_+\|_{L^1(0,\infty)} \\
&\quad + \sup_{x\in\mathbb{R}}\{(1+|x|)^{5/4}(|u_0(x)|+|u'_0(x)|)\} \\
&\quad + \sup_{0\le x<\infty}\{(1+x)(|v_-(-x)|+|v_+(x)|)\} \equiv \delta^* \ll 1.
\end{aligned}
$$

Then for all $x \in \mathbb{R}$, $t \geq 0$, the solution u to (2.1), (2.2) has the property

$$u(x,t) = \sum_{i=1}^{s} r_i \theta_i(x,t) + v(x,t),$$

(2.25)
$$v(x,t) = O(1)\delta^* \Phi(x,t), \qquad v_x(x,t) = O(1)\delta^* (t+1)^{-1/2} \Phi(x,t),$$

$$\Phi(x,t) = \sum_{i=1}^{s} [\psi_{3/2}(x,t;\lambda_i) + \phi_1(x,t;\lambda_i)\psi_{3/4}(x,t;\lambda_i)],$$

where θ_i, $i = 1, \ldots, s$, is the self-similar solution to (2.11), (2.12), given by (2.15), (2.16) or (2.18); λ_i, r_i and l_i, $i = 1, \ldots, s$, are eigenvalues and matrices consisting of right and left eigenvectors of $f'(0)$, given by (2.5) and (2.6); ψ_α and ϕ_α are defined in (2.19).

Remark 2.4. From (2.25), (2.19) we see that for $k = 0, 1$,

(2.26)
$$\left\| \frac{\partial^k}{\partial x^k} v(\cdot, t) \right\|_{L^1} = O(1)\delta^* (t+1)^{-\frac{1}{4} - \frac{k}{2}},$$

$$\left\| \frac{\partial^k}{\partial x^k} v(\cdot, t) \right\|_{L^2} = O(1)\delta^* (t+1)^{-\frac{1}{2} - \frac{k}{2}},$$

$$\left\| \frac{\partial^k}{\partial x^k} v(\cdot, t) \right\|_{L^\infty} = O(1)\delta^* (t+1)^{-\frac{3}{4} - \frac{k}{2}}.$$

All these rates are optimal.

Remark 2.5. If the left eigenspace of B associated with the zero eigenvalue is independent of u (including B being nonsingular), to obtain the pointwise estimate for $v(x,t)$ in (2.25), it is sufficient to require $u_0 \in H^4(\mathbb{R})$ rather than $u_0 \in H^8(\mathbb{R})$.

From (2.25) we see that the remainder $v(x,t)$ has decay rate $|x|^{-5/4}$ as $x \to \infty$, which is consistent with the decay rate of the initial data in (2.22), and can be improved if the decay rate of the initial data is improved. For example, if u_0 and u_0' decay exponentially, so do v and v_x in x. From (2.25) and (2.19) we also see that v and v_x have decay rates $(t+1)^{-3/4}$ and $(t+1)^{-5/4}$, respectively, along characteristic directions $dx/dt = \lambda_i$, $i = 1, \ldots, s$. These two rates are optimal and cannot be improved by the rapid decay of the initial data. Away from the characteristic directions v and v_x have decay rates $(t+1)^{-5/4}$ and $(t+1)^{-7/4}$ respectively, which can be improved up to $(t+1)^{-3/2}$ and $(t+1)^{-2}$. More precisely, we have the following theorem.

Theorem 2.6. *Suppose that system (2.1) satisfies Assumptions 2.1–2.3, and that the initial data satisfy the following smoothness and smallness assumptions:*

$$u_0 \in H^9(\mathbb{R}),$$

$$u_0(x) = O(1)(1 + |x|)^{-3/2}, \qquad u_0'(x) = O(1)(1 + |x|)^{-3/2},$$

(2.27) $$v_-(x) \equiv \int_{-\infty}^x u_0(y)dy \in L^1(\mathbb{R}^-), \quad v_+(x) \equiv \int_x^\infty u_0(y)dy \in L^1(\mathbb{R}^+),$$

$$\|u_0\|_{H^9} + \|v_-\|_{L^1(-\infty,0)} + \|v_+\|_{L^1(0,\infty)}$$

$$+ \sup_{x \in \mathbb{R}}\{(1 + |x|)^{3/2}(|u_0(x)| + |u_0'(x)|)\} \equiv \delta^* \ll 1.$$

Then for all $x \in \mathbb{R}$, $t \geq 0$, the solution u to (2.1), (2.2) has the property:

$$u(x,t) = \sum_{i=1}^s r_i[\theta_i(x,t) + v_i(x,t)],$$

(2.28) $$v_i(x,t) = O(1)\delta^* \Phi_i(x,t), \qquad v_{ix}(x,t) = O(1)\delta^*(t+1)^{-1/2}\Phi_i(x,t),$$

$$\Phi_i(x,t) = \psi_{3/2}(x,t;\lambda_i) + \sum_{j \neq i} \tilde{\psi}(x,t;\lambda_j),$$

where θ_i, $i = 1,\ldots,s$, is the self-similar solution to (2.11), (2.12), given by (2.15), (2.16) or (2.18); λ_i, r_i and l_i, $i = 1,\ldots,s$, are eigenvalues and matrices consisting of right and left eigenvectors of $f'(0)$, given by (2.5) and (2.6); ψ_α and $\tilde{\psi}$ are defined in (2.19).

Remark 2.7. The decay rates $(t+1)^{-3/2}$ and $(t+1)^{-2}$ for v_i and v_{ix}, respectively, away from the characteristic directions are optimal. They are due to the nonlinear coupling of different families and can not be improved even if the initial data are smoother and decay faster.

Remark 2.8. Similar to Remark 2.5 to Theorem 2.3, if the left eigenspace of B associated with the zero eigenvalue is independent of u, then the stringent restriction $u_0 \in H^9(\mathbb{R})$ in Theorem 2.6 is relaxed as $u_0 \in H^4(\mathbb{R})$ to obtain the pointwise estimate for v_i in (2.28), and $u_0 \in H^6(\mathbb{R})$ to obtain the pointwise estimate for v_{ix}.

   If B is a positive definite constant matrix, e.g., $B = I$, we further have the following theorem.

Theorem 2.9. *Suppose that B is a constant matrix, there exists a symmetric positive definite matrix A_0, such that $A_0 f'(0)$ is symmetric, and $A_0 B$ is symmetric positive definite,* *and the initial data satisfy the following smallness assumptions:*

(2.29) $u_0(x) = O(1)(1 + |x|)^{-3/2}, \qquad u_0' \in L^\infty(\mathbb{R}),$

(2.30) $v_-(x) \equiv \int_{-\infty}^{x} u_0(y)dy \in L^1(\mathbb{R}^-), \quad v_+(x) \equiv \int_{x}^{\infty} u_0(y)dy \in L^1(\mathbb{R}^+),$

(2.31) $\|u_0'\|_{L^\infty} + \|v_-\|_{L^1(-\infty,0)} + \|v_+\|_{L^1(0,\infty)}$

$$+ \sup_{x\in\mathbb{R}}\{(1 + |x|)^{3/2}|u_0(x)|\} \equiv \delta^* \ll 1.$$

Then for all $x \in \mathbb{R}$, $t \geq 0$, the solution u to (2.1), (2.2) has the property:

$$u(x,t) = \sum_{i=1}^{s} r_i[\theta_i(x,t) + v_i(x,t)],$$

(2.32) $$v_i(x,t) = O(1)\delta^*[\psi_{3/2}(x,t;\lambda_i) + \sum_{j\neq i}\tilde{\psi}(x,t;\lambda_j)],$$

$$v_{ix}(x,t) = O(1)\delta^*(t+1)^{-5/4},$$

where θ_i, $i = 1,\ldots,s$, is the self-similar solution to (2.11), (2.12), given by (2.15), (2.16) or (2.18); λ_i, r_i and l_i, $i = 1,\ldots,s$, are eigenvalues and matrices consisting of right and left eigenvectors of $f'(0)$, given by (2.5) and (2.6); ψ_α and $\tilde{\psi}$ are defined in (2.19).

Throughout this paper we use C to denote a universal positive constant.

3. DIFFUSION WAVES

We now discuss some properties of the diffusion waves defined in Section 2, including their convolution with the heat kernel. We also discuss the convolution of the generalized diffusion waves with the heat kernel. These properties will play an important role when assessing the effect of the nonlinear error terms in the a priori estimates in Sections 4 and 8.

Lemma 3.1. *Let θ_i, $i = 1, \ldots, s$, be the diffusion waves given by (2.15), (2.16) or (2.18). Then for all $-\infty < x < \infty$, $t \geq 0$, we have*

(3.1)
$$\theta_i(x,t) = O(1)|\delta_i|(t+1)^{-\frac{1}{2}}e^{-y_i^2},$$
$$\theta_{ix}(x,t) = O(1)|\delta_i|(t+1)^{-1}(|y_i|+1)e^{-y_i^2},$$
$$\theta_{it}(x,t) = O(1)|\delta_i|(t+1)^{-1}(y_i^2+|y_i|+1)e^{-y_i^2},$$
$$(\theta_{it}+\lambda_i\theta_{ix})(x,t) = O(1)|\delta_i|(t+1)^{-\frac{3}{2}}(y_i^2+|y_i|+1)e^{-y_i^2},$$
$$\theta_{ixx}(x,t) = O(1)|\delta_i|(t+1)^{-\frac{3}{2}}(y_i^2+|y_i|+1)e^{-y_i^2},$$
$$y_i \equiv \frac{x-\lambda_i(t+1)}{\sqrt{4\mu_i(t+1)}}.$$

Proof. By straightforward computation, we obtain (3.1) from (2.15), (2.16) and Lemma 2.2. We now prove Lemma 2.2. To simplify the notation we suppress the subscript i. Substituting $\theta(x,t) = \frac{1}{\sqrt{t}}\chi(\frac{x-\lambda t}{\sqrt{t}})$ in (2.11), we have

$$-\frac{1}{2}(y\chi)' + \frac{1}{2}Q(\chi)' = D\chi'',$$

where y is the independent variable of χ, and

$$Q(\theta) = lf''(0)(r\theta, r\theta), \qquad D = lB(0)r.$$

Integrating this equation and using the assumption

$$\lim_{y \to -\infty} y\chi(y) = \chi'(-\infty) = 0,$$

we obtain

(3.2)
$$D\chi' = -\frac{1}{2}y\chi + \frac{1}{2}Q(\chi).$$

16

By Lemma 2.1, D has eigenvalues $\mu_k > 0$, $k = 1, \ldots, m$, associated with right and left eigenvectors p_k and q_k, satisfying

$$q_k p_l = \delta_{kl}, \qquad k, l = 1, \ldots, m.$$

Set $\mu = \max\{\mu_1, \ldots, \mu_m\}$. Substituting (2.17) in (3.2), expanding $\tilde{\chi} = \sum_{k=1}^{m} \tilde{\chi}_k p_k$, and noticing that Q is a quadratic form, we find

$$(3.3) \qquad \tilde{\chi}'_k = -\frac{y}{2}\left(\frac{1}{\mu_k} - \frac{1}{\mu}\right)\tilde{\chi}_k + e^{-\frac{y^2}{4\mu}} Q_k(\tilde{\chi}), \quad k = 1, \ldots, m,$$

where $Q_k = \frac{1}{2\mu_k} q_k Q$. We first prove that (3.3) (and hence (3.2)) has a unique solution satisfying the initial condition

$$(3.4) \qquad \tilde{\chi}(0) = w \equiv \sum_{k=1}^{m} w_k p_k$$

if $|w|$ is small. With the abbreviation $a_k \equiv \frac{1}{4}(1/\mu_k - 1/\mu) \geq 0$, (3.3), together with (3.4), is equivalent to

$$(3.5) \qquad \begin{aligned} \tilde{\chi}(y) &= \sum_{k=1}^{m}\left(e^{-\alpha_k y^2} w_k + \int_0^y e^{-\alpha_k(y^2 - \eta^2) - \eta^2/4\mu} Q_k(\tilde{\chi}(\eta))d\eta\right) p_k \\ &\equiv T(\tilde{\chi})(y). \end{aligned}$$

Clearly T is an operator from C_b (all $\mathbb{R}^m$-valued bounded continuous functions on $\mathbb{R}$ with the sup norm $\|\cdot\|_\infty$) to itself, and

$$(3.6) \qquad \begin{aligned} \|T(\tilde{\chi})\|_\infty &\leq C(|w| + \|\tilde{\chi}\|_\infty^2), \\ \|T(\tilde{\chi}) - T(\bar{\chi})\|_\infty &\leq C\|\tilde{\chi} + \bar{\chi}\|_\infty \|\tilde{\chi} - \bar{\chi}\|_\infty \end{aligned}$$

for some constant $C > 0$. Thus for small ε, if $|w| \leq \varepsilon$, T is a contraction mapping on the subspace $\{\tilde{\chi} \in C_b \mid \|\tilde{\chi}\|_\infty \leq 2C\varepsilon\}$. Here C is the same constant as in (3.6). The fixed-point theorem then implies that (3.5) has a unique solution $\tilde{\chi}$, which satisfies $\|\tilde{\chi}\|_\infty < 2C\varepsilon$ by (3.6). It is easy to check by (3.3) and (3.5) that all the derivatives of $\tilde{\chi}$ are also uniformly bounded by $O(1)\varepsilon$.

We have proved that for small w, (3.2) has a unique solution χ with $\chi(0) = w$. Notice that the mass carried by the self-similar solution θ is time invariant:

$$\int_{-\infty}^{\infty} \frac{1}{\sqrt{t}} \chi\left(\frac{x - \lambda t}{\sqrt{t}}\right) dx = \int_{-\infty}^{\infty} \chi(y) dy.$$

We consider the mapping

(3.7) $$w \to \int_{-\infty}^{\infty} \chi(y) dy.$$

From (3.2) the derivative of $\chi(y) = \chi(y; w)$ with respect to w for $w = 0$ is

$$(p_1 \dots p_m) \begin{pmatrix} e^{-\frac{y^2}{4\mu_1}} & & \\ & \ddots & \\ & & e^{-\frac{y^2}{4\mu_m}} \end{pmatrix} \begin{pmatrix} q_1 \\ \vdots \\ q_m \end{pmatrix}.$$

Thus the derivative of mapping (3.7) for $w = 0$ is similar to

$$2\sqrt{\pi} \operatorname{diag}(\sqrt{\mu_1}, \dots, \sqrt{\mu_m}),$$

which is positive definite. The inverse function theorem then implies that for a given mass δ which is sufficiently small, there exists a unique self-similar solution of (2.11), (2.12). Moreover, $\tilde{\chi}$ and all its derivatives are uniformly bounded by $O(1)|\delta|$ since $|w| = O(1)|\delta|$.

$\square$

Next we consider the convolution of diffusion waves. Without ambiguity of notation, we denote the diffusion waves in the direction λ by

(3.8) $$\theta_\alpha(x, t; \lambda, \mu) = (t+1)^{-\frac{\alpha}{2}} e^{-\frac{(x-\lambda(t+1))^2}{\mu(t+1)}},$$

where $\alpha > 0$, $\mu > 0$, and λ are constants. Notice that by completing the square, we have

(3.9)
$$\begin{aligned}
&-\frac{(x - y - \lambda(t-s))^2}{\mu(t-s)} - \frac{(y - \lambda'(s+1))^2}{\mu(s+1)} \\
&= -\frac{t+1}{\mu(t-s)(s+1)} \left[y - \frac{(s+1)(x - (\lambda - \lambda')(t-s))}{t+1} \right]^2 \\
&\quad - \frac{(x - \lambda(t-s) - \lambda'(s+1))^2}{\mu(t+1)}.
\end{aligned}$$

Lemma 3.2. *Let $\alpha \geq 0$, $\beta > 0$, $\mu > 0$ and λ be constants. Then for all $-\infty < x < \infty$, $t \geq 0$, we have*

(3.10) $$\int_0^{t/2} \int_{-\infty}^{\infty} (t-s)^{-1} (t+1-s)^{-\frac{\alpha}{2}} e^{-\frac{(x-y-\lambda(t-s))^2}{\mu(t-s)}} \theta_\beta(y, s; \lambda, \mu) \, dy \, ds$$

$$= \begin{cases} O(1)\theta_\gamma(x, t; \lambda, \mu), & \text{if } \beta \neq 3 \\ O(1)\theta_\gamma(x, t; \lambda, \mu) \log(t+2), & \text{if } \beta = 3, \end{cases}$$

where $\gamma = \alpha + \min(\beta, 3) - 1;$

(3.11) $\quad \displaystyle\int_{t/2}^{t} \int_{-\infty}^{\infty} (t-s)^{-1}(t+1-s)^{-\frac{\alpha}{2}} e^{-\frac{(x-y-\lambda(t-s))^2}{\mu(t-s)}} \theta_\beta(y, s; \lambda, \mu) dy ds$

$$= \begin{cases} O(1)\theta_\gamma(x, t; \lambda, \mu), & \text{if } \alpha \neq 1 \\ O(1)\theta_\gamma(x, t; \lambda, \mu) \log(t+2), & \text{if } \alpha = 1, \end{cases}$$

where $\gamma = \min(\alpha, 1) + \beta - 1.$

Proof. Denote the left-hand side of (3.10) as I. By (3.8) and (3.9) we have

$$I = \sqrt{\pi\mu} \int_0^{t/2} (t-s)^{-\frac{1}{2}} (t+1-s)^{-\frac{\alpha}{2}} (s+1)^{-\frac{\beta}{2}+\frac{1}{2}} (t+1)^{-\frac{1}{2}} e^{-\frac{(x-\lambda(t+1))^2}{\mu(t+1)}} ds$$

$$= O(1)\theta_{\alpha+2}(x, t; \lambda, \mu) \int_0^{t/2} (s+1)^{-\frac{\beta}{2}+\frac{1}{2}} ds.$$

Evaluating the last integral, we obtain (3.10). Equation (3.11) can be proved in a similar way, where we partition the interval of integration $[t/2, t]$ into $[t/2, t-1]$ and $[t-1, t]$. $\quad\square$

Denote the characteristic function of a set $\mathcal{D}$ as $\text{char}\{\mathcal{D}\}$.

Lemma 3.3. *Let the constants* $\alpha \geq 0$, $\beta \geq 1$, $\mu > 0$, *and* $\lambda \neq \lambda'$. *Then for any fixed* $\varepsilon > 0$, $K \geq |\lambda - \lambda'|$, *and all* $-\infty < x < \infty$, $t \geq 0$, *we have*

(3.12)

$$\int_0^t \int_{-\infty}^{\infty} (t-s)^{-1}(t+1-s)^{-\frac{\alpha}{2}} e^{-\frac{(x-y-\lambda(t-s))^2}{\mu(t-s)}} \theta_\beta(y, s; \lambda', \mu) dy ds$$

$$= O(1)\Big[\theta_\gamma(x, t; \lambda, \mu + \varepsilon) + \theta_\gamma(x, t; \lambda', \mu + \varepsilon) + |x - \lambda(t+1)|^{-\frac{\beta-1}{2}} |x - \lambda'(t+1)|^{-\frac{\alpha+1}{2}}$$

$$\cdot \text{char}\big\{\min(\lambda, \lambda')(t+1) + K\sqrt{t+1} \leq x \leq \max(\lambda, \lambda')(t+1) - K\sqrt{t+1}\big\}\Big]$$

$$+ \begin{cases} O(1)\theta_\gamma(x, t; \lambda, \mu + \varepsilon) \log(t+1), & \text{if } \beta = 3 \\ 0, & \text{otherwise} \end{cases}$$

$$+ \begin{cases} O(1)\theta_\gamma(x, t; \lambda', \mu + \varepsilon) \log(t+1), & \text{if } \alpha = 1 \\ 0, & \text{otherwise}, \end{cases}$$

where $\gamma = \min(\alpha, 1) + \min(\beta, 3) - 1.$

Proof. Denote the left-hand side of (3.12) as I. Notice that for fixed α and β, $I = I(x, t; \lambda, \lambda', \mu)$. By change of variable, we have

$$I(x, t; \lambda, \lambda', \mu) = |\lambda - \lambda'| I\left(\frac{x - \lambda(t+1)}{\lambda' - \lambda}, t; 0, 1, \frac{\mu}{(\lambda' - \lambda)^2}\right).$$

Thus it is sufficient to prove the lemma for $\lambda = 0$, $\lambda' = 1$. By (3.8) and (3.9),

$$(3.13) \qquad I = \sqrt{\pi\mu} \int_0^t (t-s)^{-\frac{1}{2}} (t+1-s)^{-\frac{\alpha}{2}} (s+1)^{-\frac{\beta}{2}+\frac{1}{2}} (t+1)^{-\frac{1}{2}} e^{-\frac{(x-(s+1))^2}{\mu(t+1)}} \, ds.$$

Case 1. $x \leq 0$. Then

$$I = O(1)(t+1)^{-1-\frac{\alpha}{2}} \int_0^{t/2} (s+1)^{-\frac{\beta}{2}+\frac{1}{2}} e^{-\frac{x^2}{\mu(t+1)}} \, ds$$

$$+ O(1)(t+1)^{-\frac{\beta}{2}} \int_{t/2}^t (t-s)^{-\frac{1}{2}} (t+1-s)^{-\frac{\alpha}{2}} e^{-\frac{x^2}{\mu(t+1)} - \frac{t+1}{4\mu}} \, ds$$

$$= O(1)\theta_\gamma(x,t;0,\mu) + \begin{cases} O(1)\theta_\gamma(x,t;0,\mu)\log(t+1), & \text{if } \beta = 3 \\ 0, & \text{otherwise}, \end{cases}$$

which is bounded by the right-hand side of (3.12).

Case 2. $0 < x \leq K\sqrt{t+1}$. If $t \geq (4K)^2$, then by (3.13),

$$I = O(1)(t+1)^{-1-\frac{\alpha}{2}} \int_0^{t/2} (s+1)^{-\frac{\beta}{2}+\frac{1}{2}} \, ds$$

$$(3.14) \qquad\qquad + O(1)(t+1)^{-\frac{\beta}{2}} \int_{t/2}^t (t-s)^{-\frac{1}{2}} (t+1-s)^{-\frac{\alpha}{2}} \, ds \, e^{-\frac{t+1}{16\mu}}$$

$$= O(1)(t+1)^{-\frac{\gamma}{2}} + \begin{cases} O(1)(t+1)^{-\frac{\gamma}{2}}\log(t+1), & \text{if } \beta = 3 \\ 0, & \text{otherwise}. \end{cases}$$

It is easy to check that (3.14) is also true for $t \leq (4K)^2$. The right-hand side of (3.14) is bounded by the right-hand side of (3.12) since $0 < x \leq K\sqrt{t}$.

Case 3. $K\sqrt{t+1} < x < t+1-K\sqrt{t+1}$. Let $\delta > 0$ be small and $C > 0$ be large. By (3.13),

$$I = O(1) \int_0^{x/C-1} (t+1)^{-1-\frac{\alpha}{2}} (s+1)^{-\frac{\beta}{2}+\frac{1}{2}} e^{-\frac{x^2}{(\mu+\varepsilon)(t+1)}} \, ds$$

$$+ O(1) \int_{x/C-1}^{x+(t-x)(1-\delta)} (t-x)^{-\frac{1}{2}-\frac{\alpha}{2}} x^{-\frac{\beta}{2}+\frac{1}{2}} (t+1)^{-\frac{1}{2}} e^{-\frac{(x-(s+1))^2}{\mu(t+1)}} \, ds$$

$$+ O(1) \int_{x+(t-x)(1-\delta)}^t (t-s)^{-\frac{1}{2}} (t+1-s)^{-\frac{\alpha}{2}} (s+1)^{-\frac{\beta}{2}+\frac{1}{2}} (t+1)^{-\frac{1}{2}} e^{-\frac{(t+1-x)^2}{(\mu+\varepsilon)(t+1)}} \, ds$$

$$= O(1)\theta_\gamma(x,t;0,\mu+\varepsilon) + O(1)(t-x)^{-\frac{1}{2}-\frac{\alpha}{2}} x^{-\frac{\beta}{2}+\frac{1}{2}} + O(1)\theta_\gamma(x,t;1,\mu+\varepsilon)$$

$$+ \begin{cases} O(1)\theta_\gamma(x,t;0,\mu+\varepsilon)\log(t+1), & \text{if } \beta = 3 \\ 0, & \text{otherwise} \end{cases}$$

$$+ \begin{cases} O(1)\theta_\gamma(x,t;1,\mu+\varepsilon)\log(t+1), & \text{if } \alpha = 1 \\ 0, & \text{otherwise}, \end{cases}$$

which is bounded by the right-hand side of (3.12).

Case 4. $|x - (t+1)| \leq K\sqrt{t+1}$. This case is similar to Case 2.

Case 5. $x > t + 1 + K\sqrt{t+1}$. Notice that

$$
e^{-\frac{(x-(s+1))^2}{\mu(t+1)}} \leq \begin{cases} O(1)e^{-\frac{(x-(t+1))^2}{\mu(t+1)} - \frac{t}{C}}, & \text{if } 0 \leq s \leq t/2 \\ e^{-\frac{(x-(t+1))^2}{\mu(t+1)}}, & \text{if } t/2 \leq s \leq t. \end{cases}
$$

This case is similar to Case 1.

$\square$

Set

(3.15)
$$
\psi_\alpha(x,t;\lambda) = [(x - \lambda(t+1))^2 + t + 1]^{-\alpha/2},
$$
$$
\phi_\alpha(x,t;\lambda) = (|x - \lambda(t+1)| + t + 1)^{-\alpha/2}.
$$

Lemma 3.4. *Let the constants* $1 \leq \alpha < 3$, μ, $\mu' > 0$ *and* $\lambda \neq \lambda'$. *Let* $k \geq 0$ *be an integer. If a function* $h(x,t)$ *satisfies*

$$
h(x,t) = O(1)\theta_\alpha(x,t;\lambda',\mu'),
$$
$$
\frac{\partial^k}{\partial x^k}h(x,t) = O(1)\theta_{\alpha+k}(x,t;\lambda',\mu'),
$$

(3.16)
$$
h_t + \lambda'h_x - \frac{\mu}{4}h_{xx} = O(1)\theta_{\alpha+2}(x,t;\lambda',\mu') + (O(1)\theta_{\alpha+1}(x,t;\lambda',\mu'))_x,
$$
$$
\frac{\partial^k}{\partial x^k}\left(h_t + \lambda'h_x - \frac{\mu}{4}h_{xx}\right) = O(1)\theta_{\alpha+k+2}(x,t;\lambda',\mu') + (O(1)\theta_{\alpha+k+1}(x,t;\lambda',\mu'))_x,
$$

then for any fixed $\varepsilon > 0$, $K \geq |\lambda - \lambda'|$, *and all* $-\infty < x < \infty$, $t \geq 0$,

(3.17)
$$
\int_0^t \int_{-\infty}^\infty (t-s)^{-\frac{1}{2}} e^{-\frac{(x-y-\lambda(t-s))^2}{\mu(t-s)}} \frac{\partial^{k+1}}{\partial y^{k+1}}h(y,s)\,dy\,ds
$$
$$
= O(1)(t+1)^{-\frac{k}{2}}\left[\psi_{\frac{\alpha+1}{2}}(x,t;\lambda) + \theta_{\min(\alpha,2)}(x,t;\lambda',\mu^* + \varepsilon)\right.
$$
$$
+ |x - \lambda(t+1)|^{-\frac{\alpha}{2}}|x - \lambda'(t+1)|^{-\frac{1}{2}}
$$
$$
\left. \cdot \text{char}\{\min(\lambda,\lambda')(t+1) + K\sqrt{t+1} \leq x \leq \max(\lambda,\lambda')(t+1) - K\sqrt{t+1}\}\right],
$$

where $\mu^* = \max(\mu,\mu')$.

Proof. Again it is sufficient to prove the lemma for $\lambda = 0$, $\lambda' = 1$. We may assume $\mu \leq \mu'$. Otherwise we can always replace μ' in (3.16) by μ. Denote the left-hand side of (3.17) as

(3.18)
$$
I = I_1 + I_2,
$$

where

$$
(3.19) \quad
\begin{aligned}
I_1 &= \int_0^{\sqrt{t}} \int_{-\infty}^{\infty} (t-s)^{-\frac{1}{2}} e^{-\frac{(x-y)^2}{\mu(t-s)}} \frac{\partial^{k+1}}{\partial y^{k+1}} h(y,s)\, dy ds, \\
I_2 &= \int_{\sqrt{t}}^{t} \int_{-\infty}^{\infty} (t-s)^{-\frac{1}{2}} e^{-\frac{(x-y)^2}{\mu(t-s)}} \frac{\partial^{k+1}}{\partial y^{k+1}} h(y,s)\, dy ds.
\end{aligned}
$$

We assume $t \geq 4$. The case that $t \leq 4$ is a consequence of Lemma 3.3. By (3.16) and (3.9),

$$
(3.20) \quad
\begin{aligned}
I_1 &= O(1) \int_0^{\sqrt{t}} \int_{-\infty}^{\infty} (t-s)^{-\frac{k}{2}-1} e^{-\frac{(x-y)^2}{(\mu+\varepsilon)(t-s)}} h(y,s)\, dy ds \\
&= O(1) \int_0^{\sqrt{t}} (t-s)^{-\frac{k}{2}-\frac{1}{2}} (t+1)^{-\frac{1}{2}} (s+1)^{-\frac{\alpha}{2}+\frac{1}{2}} e^{-\frac{(x-(s+1))^2}{(\mu'+\varepsilon)(t+1)}}\, ds,
\end{aligned}
$$

where $\varepsilon > 0$ is a small constant.

Case 1.1. $x < 0$. Then

$$
(3.21) \quad
\begin{aligned}
I_1 &= O(1) \int_0^{\sqrt{t}} (t-s)^{-\frac{k}{2}-\frac{1}{2}} (t+1)^{-\frac{1}{2}} (s+1)^{-\frac{\alpha}{2}+\frac{1}{2}} e^{-\frac{x^2}{(\mu'+\varepsilon)(t+1)}}\, ds \\
&= O(1)(t+1)^{-\frac{k}{2}-\frac{\alpha}{4}-\frac{1}{4}} e^{-\frac{x^2}{(\mu'+\varepsilon)(t+1)}} \\
&= O(1)(t+1)^{-\frac{k}{2}} \theta_{\frac{\alpha+1}{2}}(x,t;0,\mu'+\varepsilon).
\end{aligned}
$$

Case 1.2. $0 \leq x \leq C\sqrt{t+1}$ with large C. By (3.20),

$$
(3.22) \quad
\begin{aligned}
I_1 &= O(1) \int_0^{\sqrt{t}} (t-s)^{-\frac{k}{2}-\frac{1}{2}} (t+1)^{-\frac{1}{2}} (s+1)^{-\frac{\alpha}{2}+\frac{1}{2}}\, ds \\
&= O(1)(t+1)^{-\frac{k}{2}-\frac{\alpha}{2}-\frac{1}{4}} = O(1)(t+1)^{-\frac{k}{2}} \theta_{\frac{\alpha+1}{2}}(x,t;0,\mu'+\varepsilon).
\end{aligned}
$$

Case 1.3. $x > C\sqrt{t+1}$. By (3.20),

$$
(3.23) \quad
\begin{aligned}
I_1 &= O(1) \int_0^{\sqrt{t}} (t-s)^{-\frac{k}{2}-\frac{1}{2}} (t+1)^{-\frac{1}{2}} (s+1)^{-\frac{\alpha}{2}+\frac{1}{2}} e^{-\frac{x^2(1-2/C)^2}{(\mu'+\varepsilon)(t+1)}}\, ds \\
&= O(1)(t+1)^{-\frac{k}{2}} \theta_{\frac{\alpha+1}{2}}(x,t;0,\mu'+2\varepsilon).
\end{aligned}
$$

Notice that from (3.8) and (3.15),

$$
(3.24) \quad \theta_\alpha(x,t;\lambda,\mu) = O(1)\psi_\alpha(x,t;\lambda)
$$

for any $\alpha > 0$, $\mu > 0$ and λ. Thus (3.21)–(3.23) imply that I_1 is bounded by the right-hand side of (3.17).

Denote the solution of

(3.25)
$$\begin{cases} w_t = \dfrac{\mu}{4} w_{xx} + \sqrt{\mu\pi} \dfrac{\partial^{k+1} h}{\partial x^{k+1}} \\ w(x, t_0) = 0 \end{cases}$$

as $w(x, t; t_0)$. Duhamel's principle and (3.19) imply that $I_2 = w(x, t; \sqrt{t})$. The initial value problem (3.25) can be rewritten as

$$\begin{cases} (w + \sqrt{\mu\pi} \dfrac{\partial^k h}{\partial x^k})_t = \dfrac{\mu}{4}(w + \sqrt{\mu\pi} \dfrac{\partial^k h}{\partial x^k})_{xx} + \sqrt{\mu\pi}\tilde{h} \\ (w + \sqrt{\mu\pi} \dfrac{\partial^k h}{\partial x^k})(x, t_0) = \sqrt{\mu\pi} \dfrac{\partial^k h}{\partial x^k}(x, t_0), \end{cases}$$

where

(3.26)
$$\tilde{h}(x, t) = \dfrac{\partial^k}{\partial x^k}(h_t + h_x - \dfrac{\mu}{4} h_{xx})(x, t).$$

Hence

(3.27)
$$I_2 = -\sqrt{\mu\pi} \dfrac{\partial^k}{\partial x^k} h(x, t) + \int_{-\infty}^{\infty} (t - \sqrt{t})^{-\frac{1}{2}} e^{-\frac{(x-y)^2}{\mu(t-\sqrt{t})}} \dfrac{\partial^k}{\partial y^k} h(y, \sqrt{t}) \, dy$$
$$+ \int_{\sqrt{t}}^{t} \int_{-\infty}^{\infty} (t - s)^{-\frac{1}{2}} e^{-\frac{(x-y)^2}{\mu(t-s)}} \tilde{h}(y, s) \, dy ds.$$

Denote the first integral in (3.27) as I_{21}. From (3.16) and (3.9) we have

(3.28)
$$I_{21} = \int_{-\infty}^{\infty} (t - \sqrt{t})^{-\frac{k+1}{2}} O(1) e^{-\frac{(x-y)^2}{(\mu+\varepsilon)(t-\sqrt{t})}} h(y, \sqrt{t}) \, dy$$
$$= O(1)(t + 1)^{-\frac{k+1}{2} - \frac{\alpha}{4}} \int_{-\infty}^{\infty} e^{-\frac{(x-y)^2}{(\mu+\varepsilon)(t-\sqrt{t})} - \frac{(y-(\sqrt{t}+1))^2}{\mu'(\sqrt{t}+1)}} \, dy$$
$$= O(1)(t + 1)^{-\frac{k}{2} - \frac{\alpha+1}{4}} e^{-\frac{(x-(\sqrt{t}+1))^2}{(\mu'+\varepsilon)(t+1)}}$$
$$= O(1)(t + 1)^{-\frac{k}{2}} \theta_{\frac{\alpha+1}{2}}(x, t; 0, \mu' + 2\varepsilon),$$

where in the last step we have considered $x < 0$, $0 \le x \le C\sqrt{t+1}$ and $x > C\sqrt{t+1}$ respectively for a large C.

Denote the second integral in (3.27) as I_{22}. By (3.26) and (3.16),

$$
\begin{aligned}
I_{22} =\ & \int_{\sqrt{t}}^{t/2} \int_{-\infty}^{\infty} (t-s)^{-\frac{1}{2}-\frac{k}{2}} O(1) e^{-\frac{(x-y)^2}{(\mu+\varepsilon/2)(t-s)}} [\theta_{\alpha+2}(y,s;1,\mu') \\
& + (t-s)^{-\frac{1}{2}} \theta_{\alpha+1}(y,s;1,\mu')]\, dy ds + \int_{t/2}^{t} \int_{-\infty}^{\infty} (t-s)^{-\frac{1}{2}} e^{-\frac{(x-y)^2}{\mu(t-s)}} \\
& \cdot [O(1)\theta_{\alpha+k+2}(y,s;1,\mu') + (O(1)\theta_{\alpha+k+1}(y,s;1,\mu'))_y]\, dy ds \\
=\ & O(1)(t+1)^{-\frac{k}{2}} \int_{\sqrt{t}}^{t} \int_{-\infty}^{\infty} (t-s)^{-\frac{1}{2}} e^{-\frac{(x-y)^2}{(\mu+\varepsilon/2)(t-s)}} \theta_{\alpha+2}(y,s;1,\mu')\, dy ds \\
& + O(1)(t+1)^{-\frac{k}{2}} \int_{\sqrt{t}}^{t} \int_{-\infty}^{\infty} (t-s)^{-1} e^{-\frac{(x-y)^2}{(\mu+\varepsilon/2)(t-s)}} \theta_{\alpha+1}(y,s;1,\mu')\, dy ds \\
\equiv\ & I_{22}^{(1)} + I_{22}^{(2)}.
\end{aligned}
$$

(3.29)

We now estimate the first term on the right-hand side. By (3.8) and (3.9),

$$
I_{22}^{(1)} = O(1)(t+1)^{-\frac{k}{2}-\frac{1}{2}} \int_{\sqrt{t}}^{t} (s+1)^{-\frac{\alpha}{2}-\frac{1}{2}} e^{-\frac{(x-(s+1))^2}{(\mu'+\varepsilon/2)(t+1)}}\, ds.
$$

Case 2.1. $x < 0$. Then

$$
\begin{aligned}
I_{22}^{(1)} &= O(1)(t+1)^{-\frac{k}{2}-\frac{\alpha+1}{4}} \int_{\sqrt{t}}^{t} (t+1)^{-\frac{1}{2}} e^{-\frac{(s+1)^2}{(\mu'+\varepsilon)(t+1)}}\, ds\ e^{-\frac{x^2}{(\mu'+\varepsilon)(t+1)}} \\
&= O(1)(t+1)^{-\frac{k}{2}} \theta_{\frac{\alpha+1}{2}}(x,t;0,\mu'+\varepsilon).
\end{aligned}
$$

(3.30)

Case 2.2. $0 \le x \le C\sqrt{t+1}$ with large C. Then

$$
\begin{aligned}
I_{22}^{(1)} &= O(1)(t+1)^{-\frac{k}{2}-\frac{\alpha+1}{4}} \int_{\sqrt{t}}^{t} (t+1)^{-\frac{1}{2}} e^{-\frac{(x-(s+1))^2}{(\mu'+\varepsilon)(t+1)}}\, ds \\
&= O(1)(t+1)^{-\frac{k}{2}-\frac{\alpha+1}{4}} = O(1)(t+1)^{-\frac{k}{2}} \theta_{\frac{\alpha+1}{2}}(x,t;0,\mu'+\varepsilon).
\end{aligned}
$$

(3.31)

Case 2.3. $x > C\sqrt{t+1}$. Then

$$
\begin{aligned}
I_{22}^{(1)} &= O(1)(t+1)^{-\frac{k}{2}-\frac{1}{2}} \left[\int_{\sqrt{t}}^{2x/C-1} (s+1)^{-\frac{\alpha}{2}-\frac{1}{2}} e^{-\frac{x^2}{(\mu'+\varepsilon)(t+1)}}\, ds \right. \\
&\quad \left. + \int_{2x/C-1}^{t} x^{-\frac{\alpha}{2}-\frac{1}{2}} e^{-\frac{(x-(s+1))^2}{(\mu'+\varepsilon)(t+1)}}\, ds \right] \\
&= O(1)(t+1)^{-\frac{k}{2}} \left[(t+1)^{-\frac{\alpha+1}{4}} \frac{x}{\sqrt{t+1}} e^{-\frac{x^2}{(\mu'+\varepsilon)(t+1)}} + x^{-\frac{\alpha+1}{2}} \right] \\
&= O(1)(t+1)^{-\frac{k}{2}} \psi_{\frac{\alpha+1}{2}}(x,t;0),
\end{aligned}
$$

(3.32)

where the second integral is zero if $2x/C - 1 \geq t$, and we have used (3.24).

Equations (3.30)–(3.32) imply that $I_{22}^{(1)}$ is bounded by the right-hand side of (3.17). As for the second term $I_{22}^{(2)}$ on the right-hand side of (3.29), we apply Lemma 3.3 to obtain the desired result. Together with (3.27), (3.16) and (3.28), I_2 is bounded by the right-hand side of (3.17). $\qquad \square$

Next we consider the evolution of the heat kernel with generalizd diffusion waves.

Lemma 3.5. *Let $\alpha \geq 0$, $\beta \geq 0$, $\mu > 0$ and λ be constants. Then for all $-\infty < x < \infty$, $t \geq 0$, we have*

$$(3.33) \quad \int_0^t \int_{-\infty}^{\infty} (t-s)^{-1}(t+1-s)^{-\frac{\alpha}{2}} e^{-\frac{(x-y-\lambda(t-s))^2}{\mu(t-s)}} (s+1)^{-\frac{\beta}{2}} \psi_{3/2}(y,s;\lambda)\, dy\, ds$$

$$= \begin{cases} O(1)(t+1)^{-\frac{\gamma}{2}} \psi_{3/2}(x,t;\lambda)\log(t+2), & \text{if } \alpha = 1 \text{ or } \beta = 3/2 \\ O(1)(t+1)^{-\frac{\gamma}{2}} \psi_{3/2}(x,t;\lambda), & \text{otherwise,} \end{cases}$$

where $\gamma = \min(\alpha,1) + \min(\beta,3/2) - 1$.

Proof. Without loss of generality we may assume $\lambda = 0$. Denote the left-hand side of (3.33) as I. Then

$$I = I_1 + I_2,$$

where

$$I_1 = O(1)\int_0^t \int_{|y| \leq \sqrt{s+1}} (t-s)^{-1}(t+1-s)^{-\frac{\alpha}{2}} e^{-\frac{(x-y)^2}{\mu(t-s)}} (s+1)^{-\frac{\beta}{2}-\frac{3}{4}}\, dy\, ds$$

$$(3.34) \quad = O(1)\int_0^t \int_{|y| \leq \sqrt{s+1}} (t-s)^{-1}(t+1-s)^{-\frac{\alpha}{2}} e^{-\frac{(x-y)^2}{\mu(t-s)}} \theta_{\beta+\frac{3}{2}}(y,s;0,\mu)\, dy\, ds,$$

$$I_2 = O(1)\int_0^t \int_{|y| \geq \sqrt{s+1}} (t-s)^{-1}(t+1-s)^{-\frac{\alpha}{2}} e^{-\frac{(x-y)^2}{\mu(t-s)}} (s+1)^{-\frac{\beta}{2}}|y|^{-\frac{3}{2}}\, dy\, ds.$$

Applying Lemma 3.2 to I_1, we see that it is bounded by the right-hand side of (3.33). To estimate I_2, we consider two cases.

Case 1. $|x| < 2\sqrt{t+1}$. Then

$$I_2 = O(1) \int_0^{t/2} (t-s)^{-1}(t+1-s)^{-\frac{\alpha}{2}}(s+1)^{-\frac{\beta}{2}} \int_{\sqrt{s+1}}^{\infty} |y|^{-\frac{3}{2}} \, dy \, ds$$

$$+ O(1) \int_{t/2}^{t} (t-s)^{-1}(t+1-s)^{-\frac{\alpha}{2}}(s+1)^{-\frac{\beta}{2}-\frac{3}{4}} \int_{|y| \geq \sqrt{s+1}} e^{-\frac{(x-y)^2}{\mu(t-s)}} \, dy \, ds$$

(3.35)
$$= O(1)(t+1)^{-1-\frac{\alpha}{2}} \int_0^{t/2} (s+1)^{-\frac{\beta}{2}-\frac{1}{4}} \, ds$$

$$+ O(1)(t+1)^{-\frac{\beta}{2}-\frac{3}{4}} \int_{t/2}^{t} (t-s)^{-\frac{1}{2}}(t+1-s)^{-\frac{\alpha}{2}} \, ds$$

$$= \begin{cases} O(1)(t+1)^{-\frac{\gamma}{2}-\frac{3}{4}} \log(t+2), & \text{if } \alpha = 1 \text{ or } \beta = 3/2 \\ O(1)(t+1)^{-\frac{\gamma}{2}-\frac{3}{4}}, & \text{otherwise,} \end{cases}$$

which is bounded by the right-hand side of (3.33).

Case 2. $|x| \geq 2\sqrt{t+1}$. We consider $x \geq 2\sqrt{t+1}$. The case $x \leq -2\sqrt{t+1}$ is the same. Considering the integrals with respect to y on $(-\infty, -\sqrt{s+1}]$, $[\sqrt{s+1}, x/2]$ and $[x/2, +\infty)$ respectively in (3.34), we obtain

$$I_2 = O(1) \int_0^{t} (t-s)^{-1}(t+1-s)^{-\frac{\alpha}{2}} e^{-\frac{x^2}{4\mu(t-s)}} (s+1)^{-\frac{\beta}{2}-\frac{1}{4}} \, ds$$

$$+ O(1) \int_0^{t} (t-s)^{-\frac{1}{2}}(t+1-s)^{-\frac{\alpha}{2}}(s+1)^{-\frac{\beta}{2}} x^{-\frac{3}{2}} \, ds$$

(3.36)
$$= O(1)(t+1)^{-1-\frac{\alpha}{2}} e^{-\frac{x^2}{4\mu t}} \int_0^{t/2} (s+1)^{-\frac{\beta}{2}-\frac{1}{4}} \, ds$$

$$+ O(1)(t+1)^{-\frac{\beta}{2}-\frac{1}{4}} \int_{t/2}^{t} (t-s)^{-1}(t+1-s)^{-\frac{\alpha}{2}} e^{-\frac{x^2}{4\mu(t-s)}} \, ds$$

$$+ O(1)(t+1)^{-\frac{\gamma}{2}} x^{-\frac{3}{2}}.$$

The first term is bounded by the right-hand side of (3.33), cf. the last step in Case 1. The third term is $O(1)(t+1)^{-\gamma/2} \psi_{3/2}(x,t;0)$. To estimate the second term, set $\eta = x/\sqrt{t-s}$ and it becomes

$$O(1)(t+1)^{-\frac{\beta}{2}-\frac{1}{4}} \int_{\frac{x}{\sqrt{t/2}}}^{\infty} \frac{2\eta^{\alpha-1}}{(\eta^2+x^2)^{\alpha/2}} e^{-\frac{\eta^2}{4\mu}} \, d\eta$$

$$= O(1)(t+1)^{-\frac{\beta}{2}-\frac{1}{4}} x^{-\alpha} \int_{\frac{x}{\sqrt{t/2}}}^{\infty} \eta^{\alpha-1} e^{-\frac{\eta^2}{4\mu}} \, d\eta$$

$$= O(1)(t+1)^{-\frac{\beta}{2}-\frac{1}{4}-\frac{\alpha}{2}} e^{-\frac{x^2}{4\mu t}} = O(1)(t+1)^{-\frac{\gamma}{2}} \psi_{3/2}(x,t;0).$$

$\square$

Lemma 3.6. *Let the constants $\alpha \geq 0$, $\beta \geq 0$, $\mu > 0$, and $\lambda \neq \lambda'$. Then for any fixed $K > 2|\lambda - \lambda'|$ and all $-\infty < x < \infty$, $t \geq 0$, we have*

$$
\int_0^t \int_{-\infty}^{\infty} (t-s)^{-1}(t+1-s)^{-\frac{\alpha}{2}} e^{-\frac{(x-y-\lambda(t-s))^2}{\mu(t-s)}} (s+1)^{-\frac{\beta}{2}} \psi_{3/2}(y,s;\lambda') \, dy \, ds
$$

$$
= O(1)(t+1)^{-\frac{\gamma}{2}} [\psi_{3/2}(x,t;\lambda) + \psi_{3/2}(x,t;\lambda')]
$$

$$
+ O(1)|x - \lambda(t+1)|^{-\frac{1}{2}\min(\beta,\frac{5}{2})-\frac{1}{4}} |x - \lambda'(t+1)|^{-\frac{1}{2}\min(\alpha,1)-\frac{1}{2}}
$$

(3.37)
$$
\cdot \operatorname{char}\{\min(\lambda,\lambda')(t+1) + K\sqrt{t+1} \leq x \leq \max(\lambda,\lambda')(t+1) - K\sqrt{t+1}\}
$$

$$
+ \left\{
\begin{array}{ll}
O(1)(t+1)^{-\frac{\gamma}{2}} \log(t+1)[\psi_{3/2}(x,t;\lambda) + \psi_{3/2}(x,t;\lambda')], & \text{if } \alpha = 1 \\
O(1)(t+1)^{-\frac{\gamma}{2}} \log(t+1)\psi_{3/2}(x,t;\lambda), & \text{if } \alpha \neq 1 \text{ and } \beta = 3/2 \\
0, & \text{otherwise,}
\end{array}
\right.
$$

where $\gamma = \min(\alpha,1) + \min(\beta,3/2) - 1$.

Proof. Without loss of generality we assume $\lambda = 0$, $\lambda' = 1$. Denote the left-hand side of (3.37) as I. Then

$$
I = I_1 + I_2,
$$

where

$$
I_1 = O(1) \int_0^t \int_{|y-(s+1)| \leq \sqrt{s+1}} (t-s)^{-1}(t+1-s)^{-\frac{\alpha}{2}} e^{-\frac{(x-y)^2}{\mu(t-s)}} (s+1)^{-\frac{\beta}{2}-\frac{3}{4}} \, dy \, ds
$$

$$
= O(1) \int_0^t \int_{-\infty}^{\infty} (t-s)^{-1}(t+1-s)^{-\frac{\alpha}{2}} e^{-\frac{(x-y)^2}{\mu(t-s)}} \theta_{\beta+3/2}(y,s;1,\mu) \, dy \, ds,
$$

(3.38)
$$
I_2 = O(1) \int_0^t \int_{|y-(s+1)| \geq \sqrt{s+1}} (t-s)^{-1}(t+1-s)^{-\frac{\alpha}{2}} e^{-\frac{(x-y)^2}{\mu(t-s)}} (s+1)^{-\frac{\beta}{2}}
$$

$$
\cdot |y - (s+1)|^{-\frac{3}{2}} \, dy \, ds.
$$

Applying Lemma 3.3 to I_1, we see that it is bounded by the right-hand side of (3.37). To estimate I_2, we consider five cases.

Case 1. $x < -2\sqrt{t+1}$. In this case

$$
I_2 = O(1) \int_0^t \int_{-\infty}^{x/2} (t-s)^{-1}(t+1-s)^{-\frac{\alpha}{2}} e^{-\frac{(x-y)^2}{\mu(t-s)}} (s+1)^{-\frac{\beta}{2}} (s+1-\frac{x}{2})^{-\frac{3}{2}} \, dy \, ds
$$

$$
+ O(1) \int_0^t \int_{x/2}^{s+1-\sqrt{s+1}} (t-s)^{-1}(t+1-s)^{-\frac{\alpha}{2}} e^{-\frac{x^2}{4\mu(t-s)}} (s+1)^{-\frac{\beta}{2}} (s+1-y)^{-\frac{3}{2}} \, dy \, ds
$$

$$+O(1)\int_0^t\int_{s+1+\sqrt{s+1}}^\infty (t-s)^{-1}(t+1-s)^{-\frac{\alpha}{2}}e^{-\frac{x^2}{\mu(t-s)}}(s+1)^{-\frac{\beta}{2}}(y-(s+1))^{-\frac{3}{2}}\,dyds$$

$$=O(1)\int_0^t (t-s)^{-\frac{1}{2}}(t+1-s)^{-\frac{\alpha}{2}}(s+1)^{-\frac{\beta}{2}}|x|^{-\frac{3}{2}}\,ds$$

$$+O(1)\int_0^t (t-s)^{-1}(t+1-s)^{-\frac{\alpha}{2}}e^{-\frac{x^2}{4\mu(t-s)}}(s+1)^{-\frac{\beta}{2}-\frac{1}{4}}\,ds.$$

The right-hand side is exactly what we had in (3.36), which is bounded by the right-hand side of (3.33), hence of (3.37).

Case2. $|x|\le K\sqrt{t+1}$. From (3.38),

$$I_2=O(1)\int_0^{t/2}(t-s)^{-1}(t+1-s)^{-\frac{\alpha}{2}}(s+1)^{-\frac{\beta}{2}-\frac{1}{4}}\,ds$$

$$+O(1)\int_{t/2}^t (t-s)^{-\frac{1}{2}}(t+1-s)^{-\frac{\alpha}{2}}(s+1)^{-\frac{\beta}{2}-\frac{3}{4}}\,ds,$$

which is what we had in (3.35) and hence bounded by the right-hand side of (3.37).

Case3. $K\sqrt{t+1}<x<t+1-K\sqrt{t+1}$. From (3.38),

$$(3.39)\qquad I_2=O(1)\int_0^t\int_{y\le s+1-\sqrt{s+1}}+O(1)\int_0^t\int_{y\ge s+1+\sqrt{s+1}}\equiv I_{21}+I_{22}.$$

Here

$$I_{21}=O(1)\int_0^{x-\sqrt{x}}(t-s)^{-1}(t+1-s)^{-\frac{\alpha}{2}}e^{-\frac{(x-(s+1))^2}{\mu(t-s)}}(s+1)^{-\frac{\beta}{2}-\frac{1}{4}}\,ds$$

$$+O(1)\int_{x-\sqrt{x}}^{x+\sqrt{x}}(t-s)^{-\frac{1}{2}}(t+1-s)^{-\frac{\alpha}{2}}(s+1)^{-\frac{\beta}{2}-\frac{3}{4}}\,ds$$

$$(3.40)\qquad +O(1)\int_{x+\sqrt{x}}^t\int_{y<\frac{s+1+x}{2}}(t-s)^{-1}(t+1-s)^{-\frac{\alpha}{2}}e^{-\frac{(x-y)^2}{\mu(t-s)}}(s+1)^{-\frac{\beta}{2}}$$

$$\cdot (s+1-x)^{-\frac{3}{2}}\,dyds+O(1)\int_{x+\sqrt{x}}^t\int_{\frac{s+1+x}{2}}^{s+1-\sqrt{s+1}}(t-s)^{-1}(t+1-s)^{-\frac{\alpha}{2}}$$

$$\cdot e^{-\frac{(s+1-x)^2}{4\mu(t-s)}}(s+1)^{-\frac{\beta}{2}}(s+1-y)^{-\frac{3}{2}}\,dyds$$

$$\equiv I_{211}+I_{212}+I_{213}+I_{214}.$$

Estimate each term respectively. We have

$$I_{211}=O(1)\int_0^{x/2-1}(t-\frac{x}{2})^{-1-\frac{\alpha}{2}}e^{-\frac{x^2}{4\mu t}}(s+1)^{-\frac{\beta}{2}-\frac{1}{4}}\,ds$$

$$+O(1)\int_{x/2-1}^{x-\sqrt{x}}(t-x)^{-\frac{1}{2}-\frac{\alpha}{2}}(t-s)^{-\frac{1}{2}}e^{-\frac{(s+1-x)^2}{\mu(t-s)}}x^{-\frac{\beta}{2}-\frac{1}{4}}\,ds$$

$$=O(1)(t+1)^{-1-\frac{\alpha}{2}}e^{-\frac{x^2}{4\mu t}}\int_0^{t/2}(s+1)^{-\frac{\beta}{2}-\frac{1}{4}}\,ds+O(1)(t-x)^{-\frac{1}{2}-\frac{\alpha}{2}}x^{-\frac{\beta}{2}-\frac{1}{4}},$$

which is bounded by the right-hand side of (3.37).

$$I_{212} = O(1) \int_{x-\sqrt{x}}^{x+\sqrt{x}} (t - x - \sqrt{x})^{-\frac{1}{2}-\frac{\alpha}{2}} (x - \sqrt{x} + 1)^{-\frac{\beta}{2}-\frac{3}{4}} \, ds$$

$$= O(1)(t + 1 - x)^{-\frac{1}{2}-\frac{\alpha}{2}} x^{-\frac{\beta}{2}-\frac{1}{4}},$$

which is also bounded by the right-hand side of (3.37). For $\varepsilon > 0$ small,

$$I_{213} = O(1) \int_{x+\sqrt{x}}^{t} (t - s)^{-\frac{1}{2}} (t + 1 - s)^{-\frac{\alpha}{2}} x^{-\frac{\beta}{2}} (s + 1 - x)^{-\frac{3}{2}} \, ds$$

$$= O(1) \int_{x+\sqrt{x}}^{t-\varepsilon(t-x)} (t - x)^{-\frac{1}{2}-\frac{\alpha}{2}} x^{-\frac{\beta}{2}} (s + 1 - x)^{-\frac{3}{2}} \, ds$$

(3.41)
$$+ O(1) \int_{t-\varepsilon(t-x)}^{t} (t - s)^{-\frac{1}{2}} (t + 1 - s)^{-\frac{\alpha}{2}} x^{-\frac{\beta}{2}} (t - x)^{-\frac{3}{2}} \, ds$$

$$= O(1)(t - x)^{-\frac{1}{2}-\frac{\alpha}{2}} x^{-\frac{\beta}{2}-\frac{1}{4}}$$

$$+ O(1) x^{-\frac{\beta}{2}} (t - x)^{-\frac{3}{2}} \left(\int_{t-\varepsilon(t-x)}^{t-\varepsilon} + \int_{t-\varepsilon}^{t} \right) (t - s)^{-\frac{1}{2}} (t + 1 - s)^{-\frac{\alpha}{2}} \, ds.$$

The last term is

$$O(1) x^{-\frac{\beta}{2}} (t - x)^{-\frac{3}{2}} + \begin{cases} O(1) x^{-\frac{\beta}{2}} (t - x)^{-1-\frac{1}{2}\min(\alpha,1)}, & \text{if } \alpha \neq 1 \\ O(1) x^{-\frac{\beta}{2}} (t - x)^{-\frac{3}{2}} \log(t - x), & \text{if } \alpha = 1. \end{cases}$$

Notice $(t - x)^{-1/2} = O(1) x^{-1/4}$. Then the above expression can be rewritten as

$$O(1) x^{-\frac{\beta}{2}-\frac{1}{4}} (t - x)^{-\frac{1}{2}\min(\alpha,1)-\frac{1}{2}}$$

$$+ \begin{cases} O(1) t^{-\frac{\gamma}{2}} \log t [\psi_{3/2}(x, t; 0) + \psi_{3/2}(x, t; 1)], & \text{if } \alpha = 1 \\ 0, & \text{if } \alpha \neq 1. \end{cases}$$

Thus the right-hand side of (3.41) is bounded by the right-hand side of (3.37).

$$I_{214} = O(1) \int_{x+\sqrt{x}}^{t} (t - s)^{-1} (t + 1 - s)^{-\frac{\alpha}{2}} e^{-\frac{(s+1-x)^2}{4\mu(t-s)}} (s + 1)^{-\frac{\beta}{2}-\frac{1}{4}} \, ds$$

$$= O(1) x^{-\frac{\beta}{2}-\frac{1}{4}} (t - x)^{-\frac{1}{2}-\frac{\alpha}{2}} \int_{x+\sqrt{x}}^{t-\varepsilon(t-x)} (t - s)^{-\frac{1}{2}} e^{-\frac{(s+1-x)^2}{4\mu(t-s)}} \, ds$$

(3.42)
$$+ O(1) x^{-\frac{\beta}{2}-\frac{1}{4}} \int_{t-\varepsilon(t-x)}^{t} (s + 1 - x)^{-2} (t + 1 - s)^{-\frac{\alpha}{2}} \, ds$$

$$= O(1) x^{-\frac{\beta}{2}-\frac{1}{4}} (t - x)^{-\frac{1}{2}-\frac{\alpha}{2}} \int_{0}^{\infty} e^{-\frac{\eta^2}{4\mu}} \, d\eta$$

$$+ O(1) x^{-\frac{\beta}{2}-\frac{1}{4}} (t - x)^{-2} \int_{t-\varepsilon(t-x)}^{t} (t + 1 - s)^{-\frac{\alpha}{2}} \, ds$$

$$= O(1) x^{-\frac{\beta}{2}-\frac{1}{4}} (t - x)^{-\frac{1}{2}-\frac{1}{2}\min(\alpha,1)}.$$

Thus we have estimate I_{21}. In (3.39), I_{22} can be estimated in a similar way:

$$I_{22} = O(1) \int_0^{x-\sqrt{x}} \int_{s+1+\sqrt{s+1}}^{\frac{s+1+x}{2}} (t-s)^{-1}(t+1-s)^{-\frac{\alpha}{2}} e^{-\frac{(x-(s+1))^2}{4\mu(t-s)}} (s+1)^{-\frac{\beta}{2}}$$

$$\cdot (y-(s+1))^{-\frac{3}{2}} \, dyds + O(1) \int_0^{x-\sqrt{x}} \int_{y \geq \frac{s+1+x}{2}} (t-s)^{-1}(t+1-s)^{-\frac{\alpha}{2}}$$

$$\cdot e^{-\frac{(x-y)^2}{\mu(t-s)}} (s+1)^{-\frac{\beta}{2}} (x-(s+1))^{-\frac{3}{2}} \, dyds + I_{212}$$

$$+ O(1) \int_{x+\sqrt{x}}^{t} (t-s)^{-1}(t+1-s)^{-\frac{\alpha}{2}} e^{-\frac{(s+1-x)^2}{\mu(t-s)}} (s+1)^{-\frac{\beta}{2}-\frac{1}{4}} \, ds$$

$$\equiv I_{221} + I_{222} + I_{212} + I_{214},$$

where

$$I_{221} = O(1) \int_0^{x-\sqrt{x}} (t-s)^{-1}(t+1-s)^{-\frac{\alpha}{2}} e^{-\frac{(x-(s+1))^2}{4\mu(t-s)}} (s+1)^{-\frac{\beta}{2}-\frac{1}{4}} \, ds,$$

which is I_{211} in (3.40);

$$I_{222} = O(1) \int_0^{x-\sqrt{x}} (t-s)^{-\frac{1}{2}}(t+1-s)^{-\frac{\alpha}{2}} (s+1)^{-\frac{\beta}{2}} (x-(s+1))^{-\frac{3}{2}} \, ds$$

$$= O(1)(t+1-x)^{-\frac{1}{2}-\frac{\alpha}{2}} \left[\int_0^{x/2} (s+1)^{-\frac{\beta}{2}} x^{-\frac{3}{2}} \, ds + \int_{x/2}^{x-\sqrt{x}} x^{-\frac{\beta}{2}} (x-(s+1))^{-\frac{3}{2}} \, ds \right]$$

$$= O(1)(t+1-x)^{-\frac{1}{2}-\frac{\alpha}{2}} x^{-\frac{1}{2}\min(\beta,\frac{5}{2})-\frac{1}{4}},$$

which is bounded by the right-hand side of (3.37); I_{212} is the same as in (3.40), and I_{214} is in (3.42).

Case 4. $|x-(t+1)| \leq K\sqrt{t+1}$. If $\beta \neq 3/2$, this case is the same as Case 2. If $\beta = 3/2$, we divide I_2 into two parts. One is the integral with respect to s on $[0, t/2]$, and the other is on $[t/2, t]$. Clearly the second part is bounded by the right-hand side of (3.37). We consider the first part,

$$O(1) \int_0^{t/2} \left(\int_{y \leq s+1-\sqrt{s+1}} + \int_{s+1+\sqrt{s+1}}^{\frac{1}{2}(s+1+x)} \right) (t-s)^{-1}(t+1-s)^{-\frac{\alpha}{2}} e^{-\frac{(x-(s+1))^2}{4\mu(t-s)}}$$

$$\cdot (s+1)^{-\frac{3}{4}} |y-(s+1)|^{-\frac{3}{2}} \, dyds$$

$$+ O(1) \int_0^{t/2} \int_{y \geq \frac{1}{2}(s+1+x)} (t-s)^{-1}(t+1-s)^{-\frac{\alpha}{2}} e^{-\frac{(x-y)^2}{\mu(t-s)}} (s+1)^{-\frac{3}{4}} (x-(s+1))^{-\frac{3}{2}} \, dyds$$

$$= O(1) \int_0^{t/2} (t-s)^{-1}(t+1-s)^{-\frac{\alpha}{2}} e^{-\frac{t}{64\mu}}(s+1)^{-1} \, ds$$

$$+ O(1) \int_0^{t/2} (t-s)^{-\frac{1}{2}}(t+1-s)^{-\frac{\alpha}{2}}(s+1)^{-\frac{3}{4}}(x-(s+1))^{-\frac{3}{2}} \, ds$$

$$= O(1)(t+1)^{-\frac{\alpha}{2}-\frac{7}{4}},$$

which is bounded by the right-hand side of (3.37).

Case 5. $x > t + 1 + K\sqrt{t+1}$. In this case,

$$I_2 = O(1) \int_0^t \left(\int_{y \le s+1-\sqrt{s+1}} + \int_{s+1+\sqrt{s+1}}^{\frac{1}{2}(s+1+x)} \right) (t-s)^{-1}(t+1-s)^{-\frac{\alpha}{2}}$$

$$\cdot e^{-\frac{(x-(s+1))^2}{4\mu(t-s)}}(s+1)^{-\frac{\beta}{2}}|y-(s+1)|^{-\frac{3}{2}} \, dy \, ds + O(1) \int_0^t \int_{y \ge \frac{1}{2}(s+1+x)} (t-s)^{-1}$$

$$\cdot (t+1-s)^{-\frac{\alpha}{2}} e^{-\frac{(x-y)^2}{\mu(t-s)}}(s+1)^{-\frac{\beta}{2}}(x-(s+1))^{-\frac{3}{2}} \, dy \, ds$$

$$= O(1) \int_0^t (t-s)^{-1}(t+1-s)^{-\frac{\alpha}{2}} e^{-\frac{(x-(s+1))^2}{4\mu(t-s)}}(s+1)^{-\frac{\beta}{2}-\frac{1}{4}} \, ds$$

$$+ O(1) \int_0^t (t-s)^{-\frac{1}{2}}(t+1-s)^{-\frac{\alpha}{2}}(s+1)^{-\frac{\beta}{2}}(x-(s+1))^{-\frac{3}{2}} \, ds$$

$$= O(1)(t+1)^{-1-\frac{\alpha}{2}} e^{-\frac{(x-(t+1))^2}{4\mu t}} - t/C \int_0^{t/2} (s+1)^{-\frac{\beta}{2}-\frac{1}{4}} \, ds$$

$$+ O(1)(t+1)^{-\frac{\beta}{2}-\frac{1}{4}} \int_{t/2}^t (t-s)^{-1-\frac{\alpha}{2}} e^{-\frac{(x-(t+1))^2}{4\mu(t-s)}} \, ds$$

$$+ O(1)(t+1)^{-\frac{\gamma}{2}}(x-(t+1))^{-\frac{3}{2}} \begin{cases} \log(t+2), & \text{if } \alpha = 1 \\ 1, & \text{otherwise.} \end{cases}$$

The first two terms are bounded by

$$O(1)(t+1)^{-\frac{\gamma}{2}} \theta_{3/2}(x,t;1,8\mu) = O(1)(t+1)^{-\frac{\gamma}{2}} \psi_{3/2}(x,t;1).$$

Thus I_2 is bounded by the right-hand side of (3.37). $\square$

Lemma 3.7. *Let* $\alpha \ge 0$, $0 \le \beta \le 2$, $\mu > 0$, *and* λ *be constants. Then for all* $-\infty < x < \infty$,

$t \geq 0$, *we have*

(3.43)
$$\int_0^t \int_{-\infty}^\infty (t-s)^{-1}(t+1-s)^{-\frac{\alpha}{2}} e^{-\frac{(x-y-\lambda(t-s))^2}{\mu(t-s)}} (s+1)^{-\frac{\beta}{2}} \phi_1(y,s;\lambda)\psi_{3/4}(y,s;\lambda)\,dy\,ds$$

$$= \begin{cases} O(1)(t+1)^{-\frac{\gamma_1}{2}}\log(t+2)\phi_1(x,t;\lambda)\psi_{3/4}(x,t;\lambda), & if\ \alpha=1\ or\ 1 \leq \beta \leq 3/2 \\ O(1)(t+1)^{-\frac{\gamma_1}{2}}\phi_1(x,t;\lambda)\psi_{3/4}(x,t;\lambda), & otherwise \end{cases}$$

$$+ \begin{cases} O(1)(t+1)^{-\frac{\gamma_2}{2}}\log(t+2)\psi_{3/2}(x,t;\lambda), & if\ \alpha=1\ or\ \beta=3/2 \\ O(1)(t+1)^{-\frac{\gamma_2}{2}}\psi_{3/2}(x,t;\lambda), & otherwise, \end{cases}$$

where $\gamma_1 = \min(\alpha,1) + (\min(\beta,1)+\min(\beta,3/2))/2 - 1$, $\gamma_2 = \min(\alpha,1) + \min(\beta,3/2) - 1$.

Proof. Without loss of generality we assume $\lambda = 0$. Denote the left-hand side of (3.43) as I. Then

$$I = \int_0^t (t-s)^{-1}(t+1-s)^{-\frac{\alpha}{2}}(s+1)^{-\frac{\beta}{2}} \left[\left(\int_{|y| \leq \sqrt{s+1}} + \int_{\sqrt{s+1} \leq |y| \leq s+1} + \int_{|y| \geq s+1} \right) \right.$$
$$\left. e^{-\frac{(x-y)^2}{\mu(t-s)}} \phi_1(y,s;0)\psi_{3/4}(y,s;0)\,dy \right] ds$$
$$\equiv I_1 + I_2 + I_3.$$

By (3.15),

$$\phi_1(y,s;0)\psi_{3/4}(y,s;0) = O(1) \begin{cases} (s+1)^{-\frac{7}{8}}, & |y| \leq \sqrt{s+1} \\ |y|^{-\frac{3}{4}}(s+1)^{-\frac{1}{2}}, & \sqrt{s+1} \leq |y| \leq s+1 \\ |y|^{-\frac{5}{4}}, & |y| \geq s+1. \end{cases}$$

The integral I_1, corresponding to $|y| \leq \sqrt{s+1}$, is estimated in the same way as in Lemma 3.5. For $|x| \leq \sqrt{t+1}$, $I_2 + I_3$ is estimated in a similar way as in (3.35). For $\sqrt{t+1} \leq x \leq t+1$, considering $y \leq x/2$ and $y \geq x/2$, we have

$$I_2 + I_3 = O(1)\int_0^t (t-s)^{-1}(t+1-s)^{-\frac{\alpha}{2}} e^{-\frac{x^2}{4\mu(t-s)}}(s+1)^{-\frac{\beta}{2}-\frac{1}{4}}\,ds$$
$$+ O(1)\int_0^t (t-s)^{-\frac{1}{2}}(t+1-s)^{-\frac{\alpha}{2}}(s+1)^{-\frac{\beta}{2}-\sigma}x^{-\frac{5}{4}+\sigma}\,ds$$

where $0 \leq \sigma \leq 1/2$. The first term has been estimated in (3.36). In the second term, we let $\sigma = \min(1/2, 1-\beta/2)$ for $0 \leq s \leq t/2$, and $\sigma = 1/2$ for $t/2 \leq s \leq t$. Thus the second term is bounded by

(3.44)
$$O(1)(t+1)^{-\frac{1}{2}\min(\alpha,1)-\frac{\beta}{2}}x^{-\frac{3}{4}} + \begin{cases} O(1)(t+1)^{-\frac{1}{2}\min(\alpha,1)-\frac{\beta}{2}}\log(t+1)x^{-\frac{3}{4}}, & \text{if } \alpha=1 \\ O(1)(t+1)^{-\frac{1}{2}-\frac{\alpha}{2}}\log(t+2)x^{-\frac{1}{4}-\frac{\beta}{2}}, & \text{if } 1\leq\beta\leq 2 \\ 0, & \text{otherwise.} \end{cases}$$

Notice that
$$(t+1)^{-\frac{1}{2}}x^{-\frac{3}{4}} = O(1)\phi_1(x,t;0)\psi_{3/4}(x,t;0),$$

and
$$x^{\frac{1}{2}-\frac{\beta}{2}} = O(1)(t+1)^{\frac{1}{4}-\frac{\beta}{4}} \qquad \text{for } 1\leq\beta\leq 2.$$

We conclude that (3.44) is bounded by the right-hand side of (3.43). For $x \geq t+1$, $I_2 + I_3$ is estimated in a similar way as in (3.36). Similarly we can prove the case $x \leq -\sqrt{t+1}$.

$\square$

Lemma 3.8. *Let the constants $\alpha \geq 0$, $0 \leq \beta \leq 2$, $\mu > 0$, and $\lambda \neq \lambda'$. Then for any fixed $K > 2|\lambda - \lambda'|$ and all $-\infty < x < \infty$, $t \geq 0$, we have*

(3.45)
$$\int_0^t \int_{-\infty}^\infty (t-s)^{-1}(t+1-s)^{-\frac{\alpha}{2}} e^{-\frac{(x-y-\lambda(t-s))^2}{\mu(t-s)}}(s+1)^{-\frac{\beta}{2}}\phi_1(y,s;\lambda')\psi_{3/4}(y,s;\lambda')\,dyds$$
$$= O(1)(t+1)^{-\frac{\gamma_1}{2}}$$
$$\cdot \begin{cases} \phi_1(x,t;\lambda)\psi_{3/4}(x,t;\lambda) + \phi_1(x,t;\lambda')\psi_{3/4}(x,t;\lambda'), & \text{if } \alpha\neq 1,\ \beta\notin[1,\frac{3}{2}] \\ \log(t+2)\phi_1(x,t;\lambda)\psi_{3/4}(x,t;\lambda) + \phi_1(x,t;\lambda')\psi_{3/4}(x,t;\lambda'), & \text{if } \alpha\neq 1,\ \beta\in[1,\frac{3}{2}] \\ \log(t+2)[\phi_1(x,t;\lambda)\psi_{3/4}(x,t;\lambda) + \phi_1(x,t;\lambda')\psi_{3/4}(x,t;\lambda')], & \text{if } \alpha=1 \end{cases}$$
$$+ O(1)(t+1)^{-\frac{\gamma_2}{2}} \begin{cases} \psi_{3/2}(x,t;\lambda) + \psi_{3/2}(x,t;\lambda'), & \text{if } \alpha\neq 1,\ \beta\neq\frac{3}{2} \\ \log(t+2)\psi_{3/2}(x,t;\lambda) + \psi_{3/2}(x,t;\lambda'), & \text{if } \alpha\neq 1,\ \beta=\frac{3}{2} \\ \log(t+2)[\psi_{3/2}(x,t;\lambda) + \psi_{3/2}(x,t;\lambda')], & \text{if } \alpha=1 \end{cases}$$
$$+ O(1)|x-\lambda(t+1)|^{-\frac{1}{2}\min(\beta,2)-\frac{1}{4}+\varepsilon}|x-\lambda'(t+1)|^{-\frac{1}{2}\min(\alpha,1)-\frac{1}{2}}$$
$$\cdot \text{char}\{\min(\lambda,\lambda')(t+1) + K\sqrt{t+1} \leq x \leq \max(\lambda,\lambda')(t+1) - K\sqrt{t+1}\},$$

where $\gamma_1 = \min(\alpha,1) + (\min(\beta,1)+\min(\beta,3/2))/2 - 1$, $\gamma_2 = \min(\alpha,1) + \min(\beta,3/2) - 1$, and $\varepsilon > 0$ arbitrarily small.

Proof. Making use of
$$\phi_1(y,s;1)\psi_{3/4}(y,s;1) = (s+1)^{-\sigma}|y-(s+1)|^{-\frac{5}{4}+\sigma}, \qquad 0\leq\sigma\leq\frac{1}{2},$$

the proof of this lemma is totally parallel to the proof of Lemma 3.6. $\square$

Lemma 3.9. *Let $\mu > 0$, σ and λ be constants. Then for all $-\infty < x < \infty$, $t \geq 0$, we have*

$$(3.46) \qquad \int_0^t e^{-(t-s)/\mu} \psi_{3/2}(x - \sigma(t-s), s; \lambda)\, ds = O(1)\psi_{3/2}(x, t; \lambda).$$

Proof. By (3.15), the left-hand side of (3.46) is

$$I = \int_0^t e^{-(t-s)/\mu}[(x - \lambda(t+1) + (\lambda - \sigma)(t - s))^2 + s + 1]^{-3/4}\, ds.$$

If $|x - \lambda(t+1)| \leq \sqrt{t+1}$, then

$$I = \int_0^t e^{-(t-s)/\mu}O(1)(s+1)^{-3/4}\, ds = O(1)(t+1)^{-3/4} = O(1)\psi_{3/2}(x, t; \lambda).$$

If $|x - \lambda(t+1)| \geq \sqrt{t+1}$, we consider two cases:

Case (i). $(x - \lambda(t+1))(\lambda - \sigma) \geq 0$. In this case

$$I = \int_0^t e^{-(t-s)/\mu}O(1)|x - \lambda(t+1)|^{-3/2}\, ds = O(1)|x - \lambda(t+1)|^{-3/2} = O(1)\psi_{3/2}(x, t; \lambda).$$

Case (ii). $(x - \lambda(t+1))(\lambda - \sigma) < 0$. In this case

$$I = \int_0^t e^{-s/\mu}O(1)\left[\left|\frac{x - \lambda(t+1)}{\lambda - \sigma} + s\right|^2 + 1\right]^{-3/4}\, ds.$$

To bound the integrand, consider whether $|(x - \lambda(t+1))/(\lambda - \sigma) + s| \geq \frac{1}{2}|(x - \lambda(t+1))/(\lambda - \sigma)|$. Thus

$$I = O(1)\int_0^t e^{-\frac{s}{\mu}}\left|\frac{x - \lambda(t+1)}{\lambda - \sigma}\right|^{-\frac{3}{2}}\, ds + O(1)\int_0^t e^{\frac{x-\lambda(t+1)}{2\mu(\lambda-\sigma)}}\, ds$$
$$= O(1)|x - \lambda(t+1)|^{-\frac{3}{2}} = O(1)\psi_{3/2}(x, t; \lambda).$$

$\square$

Similarly, we can prove

Lemma 3.10. *Let $\mu > 0$, σ and λ be constants. Then for all $-\infty < x < \infty$, $t \geq 0$,*

$$(3.47) \qquad \int_0^t e^{-(t-s)/\mu}\phi_1(x - \sigma(t-s), s; \lambda)\psi_{3/4}(x - \sigma(t-s), s; \lambda)\, ds$$
$$= O(1)\phi_1(x, t; \lambda)\psi_{3/4}(x, t; \lambda).$$

4. Systems with Diagonalizable Linearizations

In this section we prove Theorem 2.9 for the systems whose linearizations are diagonalizable. The nondiagonalizable case will be discussed at the end of Section 8.

The Cauchy problem we are now considering is

$$
(4.1) \qquad
\begin{aligned}
u_t + f(u)_x &= Bu_{xx}, && -\infty < x < \infty, \quad t > 0, \\
u(x,0) &= u_0(x), && -\infty < x < \infty.
\end{aligned}
$$

Here B is a constant matrix, and there exists a symmetric positive definite matrix A_0, such that $A_0 f'(0)$ is symmetric, and $A_0 B$ is symmetric positive definite. Let λ_i, $i = 1, \ldots, s$, be the distinct eigenvalues of $f'(0)$ with multiplicity m_i, $\sum_{i=1}^{s} m_i = n$, and l_i and r_i be the matrices consisting of left and right eigenvectors l_{ij} and r_{ij}, respectively, of $f'(0)$:

$$
(4.2) \qquad
\begin{aligned}
l_i &= \begin{pmatrix} l_{i1} \\ \vdots \\ l_{im_i} \end{pmatrix}, \qquad r_i = (r_{i1}, \ldots, r_{im_i}), \\[1em]
f'(0) r_{ij} &= \lambda_i r_{ij}, \\
l_{ij} f'(0) &= \lambda_i l_{ij}, \\
l_{ij} r_{i'j'} &= \delta_{ii'} \delta_{jj'}, \\
i, i' &= 1, \ldots, s, \quad j = 1, \ldots, m_i, \quad j' = 1, \ldots, m_{i'}.
\end{aligned}
$$

Then we further assume that B satisfies

$$
(4.3) \qquad l_i B r_j = 0 \quad \text{for } i \neq j, \quad i, j = 1, \ldots, s.
$$

For the initial data we assume that

$$
(4.4) \quad
\begin{aligned}
u_0(x) &= O(1)(1 + |x|)^{-3/2}, \qquad u_0' \in L^\infty(\mathbb{R}), \\
v_-(x) &\equiv \int_{\infty}^{x} u_0(y)\, dy \in L^1(\mathbb{R}^-), \qquad v_+(x) \equiv \int_{x}^{\infty} u_0(y) dy \in L^1(\mathbb{R}^+), \\
\|u_0'\|_{L^\infty} &+ \|v_-\|_{L^1(-\infty,0)} + \|v_+\|_{L^1(0,\infty)} + \sup_{-\infty < x < \infty} \left\{ (1 + |x|)^{3/2} |u_0(x)| \right\} \equiv \delta^* \ll 1.
\end{aligned}
$$

Decompose the solution u to (4.1) in the right eigenvector directions,

$$
(4.5) \qquad u(x,t) = \sum_{i=1}^{s} r_i u_i(x,t), \qquad u_i(x,t) = l_i u(x,t), \qquad i = 1, \ldots, s.
$$

35

Let $\theta_i(x,t)$, $i = 1, \ldots, s$, be the self-similar solution to

(4.6)
$$\theta_{it} + \lambda_i \theta_{ix} + \frac{1}{2} l_i f''(0)(r_i \theta_i, r_i \theta_i)_x = l_i B r_i \theta_{ixx},$$
$$\int_{-\infty}^{\infty} \theta_i(x,t)\, dx = \int_{-\infty}^{\infty} l_i u_0(x)\, dx \equiv \delta_i,$$

given by (2.15), (2.16) or (2.18). Our purpose here is to show that for all $-\infty < x < \infty$, $t \geq 0$, $i = 1, \ldots, s$,

(4.7)
$$v_i(x,t) \equiv u_i(x,t) - \theta_i(x,t) = O(1)\delta^* \left[\psi_{3/2}(x,t;\lambda_i) + \sum_{j \neq i} \tilde{\psi}(x,t;\lambda_j) \right],$$
$$v_{ix}(x,t) = O(1)\delta^*(t+1)^{-5/4},$$

where

(4.8)
$$\psi_\alpha(x,t;\lambda) = [(x - \lambda(t+1))^2 + t + 1]^{-\alpha/2},$$
$$\tilde{\psi}(x,t;\lambda) = [|x - \lambda(t+1)|^3 + (t+1)^2]^{-1/2}.$$

Under our assumptions, equation (2.10) satisfied by u_i takes the form

$$u_{it} + \lambda_i u_{ix} + \frac{1}{2} l_i f''(0) \left(\sum_{j=1}^{s} r_j u_j, \sum_{j=1}^{s} r_j u_j \right)_x = l_i B r_i u_{ixx} - \bar{g}_i,$$

where

$$\bar{g}_i = l_i f(u)_x - \lambda_i u_{ix} - \frac{1}{2} l_i f''(0) \left(\sum_{j=1}^{s} r_j u_j, \sum_{j=1}^{s} r_j u_j \right)_x$$
$$= (O(1)|u|^3)_x = O(1)|u|^2 |u_x|.$$

Thus with (4.6) v_i satisfies

(4.9)
$$v_{it} + \lambda_i v_{ix} = l_i B r_i v_{ixx} + g_{ix}$$
$$v_{ixt} + \lambda_i v_{ixx} = l_i B r_i v_{ixxx} + \tilde{g}_{ix},$$

where

(4.10) $g_i = -\dfrac{1}{2} l_i \displaystyle\sum_{j \neq i} f''(0)(r_j \theta_j, r_j \theta_j) - l_i \sum_{j=1}^{s} f''(0)(r_j \theta_j, r_j v_j)$

$$+ O(1) \sum_{j \neq k} |\theta_j|(|v_k| + |\theta_k|) + O(1) \sum_{j=1}^{s} |v_j|^2 + O(1) \sum_{j=1}^{s} |\theta_j|^3,$$

$$\tilde{g}_i = g_{ix} = -\frac{1}{2} l_i \sum_{j \neq i} f''(0)(r_j \theta_j, r_j \theta_j)_x$$

$$+ O(1) \sum_{j,k=1}^{s} (|\theta_j||v_{kx}| + |\theta_{jx}||v_k| + |\theta_j|^2|\theta_{kx}| + |v_j||v_{kx}|) + O(1) \sum_{j \neq k} |\theta_{jx}||\theta_k|.$$

By Lemma 2.1 we may assume

(4.11)
$$l_i B r_i = \mathrm{diag}(\mu_{i1}, \ldots, \mu_{im_i})$$

with $\mu_{ij} > 0$, $j = 1, \ldots, m_i$, since for each $1 \le i \le s$ we can perform the same linear transform to v_i and θ_i in (4.9) and (4.10) to diagonalize $l_i B r_i$. Let

(4.12)
$$v_i = (v_{i1}, \ldots, v_{im_i})^t.$$

Duhamel's principle and (4.9) imply that for $1 \le j \le m_i$,

(4.13)
$$v_{ij}(x, t) = I + II + III + IV + V + VI,$$

with

(4.14)
$$I = \int_{-\infty}^{\infty} \frac{1}{\sqrt{4\pi\mu_{ij}t}} e^{-\frac{(x-y-\lambda_i t)^2}{4\mu_{ij}t}} v_{ij}(y, 0)\, dy,$$

$$II = -\frac{1}{2} \int_0^t \int_{-\infty}^{\infty} \frac{1}{\sqrt{4\pi\mu_{ij}(t-\tau)}} e^{-\frac{(x-y-\lambda_i(t-\tau))^2}{4\mu_{ij}(t-\tau)}} \left[l_{ij} \sum_{k \ne i} f''(0)(r_k\theta_k, r_k\theta_k)(y, \tau) \right]_y dy d\tau,$$

$$III = -\int_0^t \int_{-\infty}^{\infty} \frac{1}{\sqrt{4\pi\mu_{ij}(t-\tau)}} e^{-\frac{(x-y-\lambda_i(t-\tau))^2}{4\mu_{ij}(t-\tau)}} \left[l_{ij} \sum_{k=1}^{s} f''(0)(r_k\theta_k, r_k v_k)(y, \tau) \right]_y dy d\tau,$$

$$IV = \int_0^t \int_{-\infty}^{\infty} \frac{1}{\sqrt{4\pi\mu_{ij}(t-\tau)}} e^{-\frac{(x-y-\lambda_i(t-\tau))^2}{4\mu_{ij}(t-\tau)}} \left[O(1) \sum_{k \ne k'} |\theta_k|(|v_{k'}| + |\theta_{k'}|)(y, \tau) \right]_y dy d\tau,$$

$$V = \int_0^t \int_{-\infty}^{\infty} \frac{1}{\sqrt{4\pi\mu_{ij}(t-\tau)}} e^{-\frac{(x-y-\lambda_i(t-\tau))^2}{4\mu_{ij}(t-\tau)}} \left[O(1) \sum_{k=1}^{s} |v_k|^2(y, \tau) \right]_y dy d\tau,$$

$$VI = \int_0^t \int_{-\infty}^{\infty} \frac{1}{\sqrt{4\pi\mu_{ij}(t-\tau)}} e^{-\frac{(x-y-\lambda_i(t-\tau))^2}{4\mu_{ij}(t-\tau)}} \left[O(1) \sum_{k=1}^{s} |\theta_k|^3(y, \tau) \right]_y dy d\tau,$$

and

(4.15)
$$v_{ijx}(x, t) = \tilde{I} + \widetilde{II} + \widetilde{III},$$

with

(4.16)

$$\tilde{I} = \int_{-\infty}^{\infty} \frac{1}{\sqrt{4\pi\mu_{ij}t}} e^{-\frac{(x-y-\lambda_i t)^2}{4\mu_{ij}t}} v_{ijy}(y,0)\, dy,$$

$$\widetilde{II} = -\frac{1}{2}\int_0^t \int_{-\infty}^{\infty} \frac{1}{\sqrt{4\pi\mu_{ij}(t-\tau)}} e^{-\frac{(x-y-\lambda_i(t-\tau))^2}{4\mu_{ij}(t-\tau)}} \left[l_{ij} \sum_{k\neq i} f''(0)(r_k\theta_k, r_k\theta_k)(y,\tau) \right]_{yy} dy d\tau,$$

$$\widetilde{III} = \int_0^{t/2} \int_{-\infty}^{\infty} \frac{1}{\sqrt{4\pi\mu_{ij}(t-\tau)}} e^{-\frac{(x-y-\lambda_i(t-\tau))^2}{4\mu_{ij}(t-\tau)}} \left[O(1) \sum_{k,k'=1}^s |\theta_k||v_{k'}| + O(1)\sum_{k\neq k'} |\theta_k||\theta_{k'}| \right.$$

$$\left. + O(1)\sum_{k=1}^s (|v_k|^2 + |\theta_k|^3) \right]_{yy} (y,\tau)\, dy d\tau$$

$$+ \int_{t/2}^t \int_{-\infty}^{\infty} \frac{1}{\sqrt{4\pi\mu_{ij}(t-\tau)}} e^{-\frac{(x-y-\lambda_i(t-\tau))^2}{4\mu_{ij}(t-\tau)}} \left[O(1)\sum_{k,k'=1}^s (|\theta_k||v_{k'y}| + |\theta_{ky}||v_{k'}| \right.$$

$$\left. + |\theta_k|^2|\theta_{k'y}| + |v_k||v_{k'y}|) + O(1)\sum_{k\neq k'} |\theta_{ky}||\theta_{k'}| \right]_y (y,\tau)\, dy d\tau.$$

We now derive a priori estimate for v_i. Set

(4.17) $$M(t) = \sup_{0\leq\tau\leq t} \max_{1\leq i\leq s} \left\{ \left\| v_i(\cdot,\tau)[\psi_{3/2}(\cdot,\tau;\lambda_i) + \sum_{k\neq i} \tilde{\psi}(\cdot,\tau;\lambda_k)]^{-1} \right\|_{L^\infty} \right.$$
$$\left. + \|v_{ix}(\cdot,\tau)\|_{L^\infty} (\tau+1)^{5/4} \right\}.$$

Then for $-\infty < x < \infty$, $t \geq 0$, $i = 1, \ldots, s$,

$$|v_i(x,t)| \leq M(t)\left[\psi_{3/2}(x,t;\lambda_i) + \sum_{k\neq i} \tilde{\psi}(x,t;\lambda_k)\right],$$

(4.18)

$$|v_{ix}(x,t)| \leq M(t)(t+1)^{-5/4}.$$

We estimate each term in (4.13).

From (4.5), (4.4), (3.1) and (4.6) we see that

$$v_i(x,0) = l_i u_0(x) - \theta_i(x,0) = O(1)\delta^*(1+|x|)^{-3/2},$$

$$v_{ix}(x,0) = O(1)\delta^*,$$

(4.19) $$\int_{-\infty}^{\infty} |v_i(x,0)|\, dx = O(1)\delta^*, \qquad \int_{-\infty}^{\infty} v_i(x,0)\, dx = 0,$$

$$\int_{-\infty}^{\infty} |\eta_i(x)|\, dx = O(1)\delta^* \quad \text{with} \quad \eta_i(x) = \int_{-\infty}^x v_i(y,0)\, dy.$$

To estimate I we consider two cases.

Case 1. $|x - \lambda_i t| \leq (t+1)^{2/3}$. Let the jth component of η_i be η_{ij}. By (4.14),

$$
(4.20) \quad
\begin{aligned}
I &= \int_{-\infty}^{\infty} \frac{1}{\sqrt{4\pi\mu_{ij}t}} e^{-\frac{(x-y-\lambda_i t)^2}{4\mu_{ij}t}} \eta_{ij}'(y)\, dy = O(1)t^{-1} \int_{-\infty}^{\infty} |\eta_{ij}(y)|\, dy \\
&= O(1)\delta^*(t+1)^{-1} = O(1)\delta^*\tilde{\psi}(x,t;\lambda_i).
\end{aligned}
$$

Case 2. $|x-\lambda_i t| \geq (t+1)^{2/3}$. We consider $x-\lambda_i t \geq (t+1)^{2/3}$. The case $x-\lambda_i t \leq -(t+1)^{2/3}$ is similar. By (4.14) and (4.19),

$$
(4.21) \quad
\begin{aligned}
I &= \int_{-\infty}^{(x-\lambda_i t)/2} O(1)t^{-\frac{1}{2}} e^{-\frac{(x-\lambda_i t)^2}{16\mu_{ij}t}} |v_{ij}(y,0)|\, dy \\
&\quad + \int_{(x-\lambda_i t)/2}^{\infty} O(1)t^{-\frac{1}{2}} e^{-\frac{(x-y-\lambda_i t)^2}{4\mu_{ij}t}} \delta^*(1+y)^{-\frac{3}{2}}\, dy \\
&= O(1)t^{-\frac{1}{2}} e^{-\frac{t^{1/3}}{C} - \frac{(x-\lambda_i t)^2}{Ct}} \delta^* + O(1)\delta^*(x-\lambda_i t)^{-\frac{3}{2}} = O(1)\delta^*\tilde{\psi}(x,t;\lambda_i).
\end{aligned}
$$

Equations (4.20) and (4.21) imply that

$$
(4.22) \quad I = O(1)\delta^*\tilde{\psi}(x,t;\lambda_i).
$$

To estimate II we set

$$
(4.23) \quad h_k(x,t) = l_{ij}f''(0)(r_k\theta_k, r_k\theta_k)(x,t), \qquad 1 \leq k \leq s.
$$

we recall (3.8),

$$
(4.24) \quad \theta_\alpha(x,t;\lambda,\mu) = (t+1)^{-\frac{\alpha}{2}} e^{-\frac{(x-\lambda(t+1))^2}{\mu(t+1)}}.
$$

By (3.1) it is clear that

$$
(4.25) \quad h_k(x,t) = O(1)\delta^{*2}\theta_2(x,t;\lambda_k, 2\mu_k),
$$

and

$$
(4.26) \quad
\begin{aligned}
&h_{kt} + \lambda_k h_{kx} - \mu_{ij}h_{kxx} \\
&= 2l_{ij}f''(0)(r_k\theta_k, r_k(\theta_{kt} + \lambda_k\theta_{kx})) + (O(1)|\theta_k||\theta_{kx}|)_x \\
&= O(1)(|\theta_k||\theta_{kt} + \lambda_k\theta_{kx}|) + (O(1)|\theta_k||\theta_{kx}|)_x \\
&= O(1)\delta^{*2}\theta_4(x,t;\lambda_k, 2\mu_k + \varepsilon) + (O(1)\delta^{*2}\theta_3(x,t;\lambda_k, 2\mu_k + \varepsilon))_x,
\end{aligned}
$$

where $\mu_k > 0$ is the maximum eigenvalue of $l_k B r_k$, and $\varepsilon > 0$ is an arbitrarily fixed small constant. We now apply Lemma 3.4 to II in (4.14) with $\alpha = 2$, $k = 0$, $\lambda = \lambda_i$, and $\lambda' = \lambda_k$:

$$
\begin{aligned}
II &= -\frac{1}{4\sqrt{\pi \mu_{ij}}} \sum_{k \neq i} \int_0^t \int_{-\infty}^{\infty} (t-\tau)^{-\frac{1}{2}} e^{-\frac{(x-y-\lambda_i(t-\tau))^2}{4\mu_{ij}(t-\tau)}} h_{ky}(y,\tau)\, dy d\tau \\
&= \sum_{k \neq i} O(1)\delta^{*2}\big[\psi_{3/2}(x,t;\lambda_i) + \theta_2(x,t;\lambda_k,\mu^*) \\
&\quad + |x-\lambda_i(t+1)|^{-1}|x-\lambda_k(t+1)|^{-\frac{1}{2}} \\
&\quad \cdot \text{char}\{\min(\lambda_i,\lambda_k)(t+1) + K\sqrt{t+1} \le x \le \max(\lambda_i,\lambda_k)(t+1) - K\sqrt{t+1}\}\big] \\
&= O(1)\delta^{*2}\big[\psi_{3/2}(x,t;\lambda_i) + \sum_{k \neq i} \tilde{\psi}(x,t;\lambda_k)\big],
\end{aligned}
$$

(4.27)

where μ^*, $K > 0$ are constants.

To estimate III similarly we set

$$
h_k(x,t) = l_{ij} f''(0)(r_k \theta_k, r_k v_k)(x,t), \qquad 1 \le k \le s.
$$

We have by (3.1), (4.18) and (4.8),

$$
h_k(x,t) = O(1)|\theta_k||v_k| = O(1)\delta^* M(t)\theta_{5/2}(x,t;\lambda_k, 4\mu_k),
$$

and also by (4.9),

$$
\begin{aligned}
&h_{kt} + \lambda_k h_{kx} - \mu_{ij} h_{kxx} \\
&= l_{ij} f''(0)(r_k(\theta_{kt} + \lambda_k \theta_{kx}), r_k v_k) + l_{ij} f''(0)(r_k \theta_k, r_k(v_{kt} + \lambda_k v_{kx})) \\
&\quad + (O(1)|\theta_{kx}||v_k| + O(1)|\theta_k||v_{kx}|)_x \\
&= O(1)\delta^* M(t)\theta_4(x,t;\lambda_k, 4\mu_k + \varepsilon) + (O(1)\delta^* M(t)\theta_3(x,t;\lambda_k, 4\mu_k + \varepsilon))_x \\
&\quad + l_{ij} f''(0)(r_k \theta_k, r_k(l_k B r_k v_{kxx} + g_{kx})),
\end{aligned}
$$

where $\varepsilon > 0$ is small. By (4.10) the last term is

$$
\begin{aligned}
&l_{ij} f''(0)(r_k \theta_k, r_k(l_k B r_k v_{kx} + g_k))_x - l_{ij} f''(0)(r_k \theta_{kx}, r_k(l_k B r_k v_{kx} + g_k)) \\
&= (O(1)\delta^*(\delta^* + M(t) + M^2(t))\theta_3(x,t;\lambda_k, 4\mu_k))_x \\
&\quad + O(1)\delta^*(\delta^* + M(t) + M^2(t))\theta_4(x,t;\lambda_k, 4\mu_k + \varepsilon).
\end{aligned}
$$

Thus

$$h_{kt} + \lambda_k h_{kx} - \mu_{ij} h_{kxx} = O(1)\delta^*(\delta^* + M(t) + M^2(t))\theta_4(x, t; \lambda_k, 4\mu_k + \varepsilon)$$
$$+ (O(1)\delta^*(\delta^* + M(t) + M^2(t))\theta_3(x, t; \lambda_k, 4\mu_k + \varepsilon))_x.$$

Now for those terms of III in (4.14) with $k \neq i$ again we apply Lemma 3.4, while for the term $k = i$ we apply Lemma 3.2. We obtain

$$III = -\frac{1}{\sqrt{4\pi\mu_{ij}}} \int_0^t \int_{-\infty}^\infty (t-\tau)^{-\frac{1}{2}} e^{-\frac{(x-y-\lambda_i(t-\tau))^2}{4\mu_{ij}(t-\tau)}}$$
$$\cdot \left[O(1)\delta^* M(\tau)\theta_{5/2}(y, \tau; \lambda_i, 4\mu_i) + \sum_{k \neq i} h_k(y, \tau) \right]_y dy d\tau$$

(4.28)
$$= O(1)\delta^* M(t) \int_0^t \int_{-\infty}^\infty (t-\tau)^{-1} e^{-\frac{(x-y-\lambda_i(t-\tau))^2}{(4\mu_{ij}+\varepsilon)(t-\tau)}} \theta_{5/2}(y, \tau; \lambda_i, 4\mu_i) \, dy d\tau$$
$$+ O(1)\delta^*(\delta^* + M(t) + M^2(t))\left[\psi_{3/2}(x, t; \lambda_i) + \sum_{k \neq i} \tilde{\psi}(x, t; \lambda_k) \right]$$
$$= O(1)\delta^*(\delta^* + M(t) + M^2(t))\left[\psi_{3/2}(x, t; \lambda_i) + \sum_{k \neq i} \tilde{\psi}(x, t; \lambda_k) \right],$$

where in the last step we have used (3.24) with $\alpha = 3/2$.

To estimate IV and VI in (4.14) we use Lemmas 3.2 and 3.3. By (3.1), (4.18), (4.8), (4.24), (3.10)–(3.12), for small ε,

(4.29)
$$IV + VI = O(1) \int_0^t \int_{-\infty}^\infty (t-\tau)^{-1} e^{-\frac{(x-y-\lambda_i(t-\tau))^2}{(4\mu_{ij}+\varepsilon)(t-\tau)}} \delta^*(\delta^* + M(\tau))$$
$$\cdot \sum_{k=1}^s \theta_3(y, \tau; \lambda_k, 4\mu_k + \varepsilon) \, dy d\tau$$
$$= O(1)\delta^*(\delta^* + M(t))\Big\{ \theta_{2-\varepsilon}(x, t; \lambda_i, \mu^*) + \sum_{k \neq i} [\theta_2(x, t; \lambda_k, \mu^*)$$
$$+ |x - \lambda_i(t+1)|^{-1}|x - \lambda_k(t+1)|^{-\frac{1}{2}}$$
$$\cdot \text{char}\{\min(\lambda_i, \lambda_k)(t+1) + K\sqrt{t+1} \leq x \leq \max(\lambda_i, \lambda_k)(t+1) - K\sqrt{t+1}\}] \Big\}$$
$$= O(1)\delta^*(\delta^* + M(t))\left[\psi_{3/2}(x, t; \lambda_i) + \sum_{k \neq i} \tilde{\psi}(x, t; \lambda_k) \right],$$

where μ^*, $K > 0$ are constants.

To estimate V in (4.14) we use Lemmas 3.5 and 3.6. Notice that by (4.8)

$$\tilde{\psi}(x, t; \lambda) = O(1)\psi_{3/2}(x, t, \lambda).$$

By (4.18), (4.8), (3.33) and (3.37), for small ε,

(4.30)

$$
V = O(1) \int_0^t \int_{-\infty}^{\infty} (t-\tau)^{-1} e^{-\frac{(x-y-\lambda_i(t-\tau))^2}{(4\mu_{ij}+\varepsilon)(t-\tau)}} M^2(\tau)(\tau+1)^{-\frac{3}{4}} \sum_{k=1}^s \psi_{3/2}(y,\tau;\lambda_k)\, dy d\tau
$$

$$
= O(1) M^2(t) \bigg\{ (t+1)^{-\frac{1}{4}+\varepsilon} \psi_{3/2}(x,t;\lambda_i) + \sum_{k\neq i} \big[(t+1)^{-\frac{1}{4}} \psi_{3/2}(x,t;\lambda_k)
$$

$$
+ |x-\lambda_i(t+1)|^{-1} |x-\lambda_k(t+1)|^{-\frac{1}{2}}
$$

$$
\cdot \operatorname{char}\{ \min(\lambda_i,\lambda_k)(t+1) + K\sqrt{t+1} \leq x \leq \max(\lambda_i,\lambda_k)(t+1) - K\sqrt{t+1}\} \big] \bigg\}
$$

$$
= O(1) M^2(t) \big[\psi_{3/2}(x,t;\lambda_i) + \sum_{k\neq i} \tilde{\psi}(x,t;\lambda_k) \big],
$$

where $K > 0$ is a constant.

Equations (4.13), (4.22), (4.27)–(4.30) imply that for all $-\infty < x < \infty$, $t \geq 0$, $i = 1$, $\dots, s$,

(4.31) $$ |v_i(x,t)| \leq C(\delta^* + \delta^* M(t) + M^2(t)) \big[\psi_{3/2}(x,t;\lambda_i) + \sum_{k\neq i} \tilde{\psi}(x,t;\lambda_k) \big]. $$

Next we estimate v_{ix}. By (4.19), $\tilde{I}$ in (4.16) is

(4.32) $$ \tilde{I} = O(1) \int_{-\infty}^{\infty} t^{-\frac{3}{2}} e^{-\frac{(x-y-\lambda_i t)^2}{(4\mu_{ij}+\varepsilon)t}} \eta_{ij}(y)\, dy = O(1)(t+1)^{-\frac{3}{2}} \delta^* $$

for $t \geq 1$, which is obviously also true for $t < 1$. For $\widetilde{II}$ in (4.16), let h_k be defined by (4.23). Then besides (4.25) and (4.26) we have by (3.1)

$$
h_{kx}(x,t) = O(1)|\theta_{kx}||\theta_k| = O(1)\delta^{*2}\theta_3(x,t;\lambda_k,2\mu_k+\varepsilon),
$$

$$
(h_{kt} + \lambda_k h_{kx} - \mu_{ij} h_{kxx})_x
$$

$$
= \frac{\partial}{\partial x} \big[2l_{ij} f''(0)(r_k\theta_k, r_k(\theta_{kt}+\lambda_k\theta_{kx})) + O(1)(|\theta_k||\theta_{kxx}| + |\theta_{kx}|^2) \big]
$$

$$
= (O(1)\delta^{*2}\theta_4(x,t;\lambda_k,2\mu_k+\varepsilon))_x.
$$

Apply Lemma 3.4 to $\widetilde{II}$. Similar to (4.27) we obtain

$$
\widetilde{II} = -\frac{1}{4\sqrt{\pi\mu_{ij}}} \sum_{k\neq i} \int_0^t \int_{-\infty}^{\infty} (t-\tau)^{-\frac{1}{2}} e^{-\frac{(x-y-\lambda_i(t-\tau))^2}{4\mu_{ij}(t-\tau)}} h_{kyy}(y,\tau)\, dy d\tau
$$

(4.33) $$ = O(1)\delta^{*2}(t+1)^{-\frac{1}{2}} \big[\psi_{3/2}(x,t;\lambda_i) + \sum_{k\neq i} \tilde{\psi}(x,t;\lambda_k) \big] $$

$$
= O(1)\delta^{*2}(t+1)^{-\frac{5}{4}},
$$

using (4.8). By (3.1) and (4.18), rewrite $\widetilde{III}$ in (4.16) as

$$
\widetilde{III} = O(1) \int_0^{t/2} \int_{-\infty}^{\infty} (t-\tau)^{-\frac{3}{2}} e^{-\frac{(x-y-\lambda_i(t-\tau))^2}{(4\mu_{ij}+\varepsilon)(t-\tau)}} \sum_{k=1}^{s} \big[(\delta^* M(\tau) + \delta^{*2})\theta_{5/2}(y,\tau;\lambda_k,4\mu_k)
$$
$$
+ M^2(\tau)(1+\tau)^{-\frac{3}{4}} \psi_{3/2}(y,\tau;\lambda_k)\big] \, dy d\tau
$$
$$
+ O(1) \int_{t/2}^{t} \int_{-\infty}^{\infty} (t-\tau)^{-1} e^{-\frac{(x-y-\lambda_i(t-\tau))^2}{(4\mu_{ij}+\varepsilon)(t-\tau)}} \sum_{k=1}^{s} \big[(\delta^* M(\tau) + \delta^{*2})\theta_{7/2}(y,\tau;\lambda_k,4\mu_k+\varepsilon)
$$
$$
+ M^2(\tau)(1+\tau)^{-\frac{5}{4}} \psi_{3/2}(y,\tau;\lambda_k)\big] \, dy d\tau
$$
$$
= O(1)(\delta^{*2} + \delta^* M(t) + M^2(t))(t+1)^{-\frac{1}{2}} \int_0^{t} \int_{-\infty}^{\infty} (t-\tau)^{-1} e^{-\frac{(x-y-\lambda_i(t-\tau))^2}{(4\mu_{ij}+\varepsilon)(t-\tau)}}
$$
$$
\cdot \sum_{k=1}^{s} \big[\theta_{5/2}(y,\tau;\lambda_k,4\mu_k+\varepsilon) + (1+\tau)^{-\frac{3}{4}} \psi_{3/2}(y,\tau;\lambda_k)\big] \, dy d\tau.
$$

Apply Lemmas 3.2, 3.3, 3.5 and 3.6 to the right-hand side. We obtain

$$(4.34) \qquad \widetilde{III} = O(1)(\delta^{*2} + \delta^* M(t) + M^2(t))(t+1)^{-\frac{5}{4}}.$$

Equations (4.15), (4.32)–(4.34) together imply that for $-\infty < x < \infty$, $t \geq 0$, $i = 1, \dots, s$,

$$(4.35) \qquad |v_{ix}(x,t)| \leq C(\delta^* + \delta^* M(t) + M^2(t))(t+1)^{-\frac{5}{4}}.$$

By the definition of $M(t)$ in (4.17), inequalities (4.31) and (4.35) imply that

$$M(t) \leq C(\delta^* + \delta^* M(t) + M^2(t)).$$

If $M(t)$ and δ^* are small, we have $M(t) \leq C\delta^*$. By continuity, $M(t) \leq C\delta^*$ if δ^* is small. With (4.18) we thus have proved (4.7).

5. Green's Function for a 2 × 2 Linear System

In this and next sections we study the Green's function for the linearization of the viscous conservation laws (2.1). In this section we explore a typical 2 × 2 system, the p-system. This simplest model provides us a clear picture how the behavior of the eigenvalues of $-f'(0) + \xi B(0)$ for small ξ and for large ξ plays an important role in Green's function. Most estimates in this section apply to general $n \times n$ systems.

The nonlinear p-system describes the compressible, isentropic, viscous 1-D flow in Lagrangean coordinates. The linearization of the p-system is

$$(5.1) \qquad \begin{cases} w_{1t} - w_{2x} = 0 \\ w_{2t} + p'w_{1x} = \mu w_{2xx}, \qquad -\infty < x < \infty, \quad t > 0, \end{cases}$$

where $p' < 0$ and $\mu > 0$ are constants. Physically w_1 and w_2 represent the perturbations of specific volume and velocity about a constant state, respectively, while p' is the derivative of pressure and μ the viscosity at the constant state. Set

$$(5.2) \qquad u = \begin{pmatrix} w_1 \\ w_2 \end{pmatrix}, \qquad A = \begin{pmatrix} 0 & -1 \\ p' & 0 \end{pmatrix}, \qquad B = \begin{pmatrix} 0 & 0 \\ 0 & \mu \end{pmatrix}.$$

Equation (5.1) can be written as

$$(5.3) \qquad u_t + Au_x = Bu_{xx}.$$

The Green's function $G(x,t)$ of (5.1) or (5.3) is the solution matrix to (5.3) satisfying the initial condition

$$(5.4) \qquad G(x,0) = \delta(x)I,$$

where $\delta(x)$ is the Dirac δ-function and I is the 2 × 2 identity matrix.

Our purpose is to obtain a pointwise estimate for G. Here the difficulty lies in the fact that B is singular and hence system (5.3) is not diagonalizable. To cope with the difficulty, we approximate (5.3) by a diagonalizable system, whose Green's function is a combination of heat kernels, and then estimate the remainder by studying the Fourier and inverse Fourier transforms. The result in this section is due to Zeng [Ze]. Here, however, we give a simpler proof, which can be extended to $n \times n$ systems in the next section.

As usual, we use the Fourier transform

$$\hat{g}(\xi, t) = \int_{-\infty}^{\infty} g(x, t) e^{-ix\xi} \, dx$$

for the estimates of the Green's function. Since G satisfies (5.3), (5.4), we have

(5.5)
$$\begin{cases} \hat{G}_t = i\xi E(i\xi)\hat{G} \\ \hat{G}(\xi, 0) = I, \end{cases}$$

where

(5.6)
$$E(z) = -A + zB, \quad z = i\xi.$$

With (5.2), it is straightforward to compute the spectral representation

(5.7)
$$E(z) = \tilde{\lambda}_-(z)\tilde{P}_-(z) + \tilde{\lambda}_+(z)\tilde{P}_+(z),$$

where $\tilde{\lambda}_\pm(z)$ are eigenvalues of $E(z)$, and $\tilde{P}_\pm(z)$ are corresponding eigenprojections, satisfying

(5.8)
$$\tilde{P}_- + \tilde{P}_+ = I, \qquad \tilde{P}_\pm^2 = \tilde{P}_\pm, \qquad \tilde{P}_-\tilde{P}_+ = \tilde{P}_+\tilde{P}_- = 0.$$

In fact, let $l_\pm(z)$ and $r_\pm(z)$ be the left and right eigenvectors satisfying

$$l_\pm(z)E(z) = \tilde{\lambda}_\pm(z)l_\pm(z), \qquad E(z)r_\pm(z) = \tilde{\lambda}_\pm(z)r_\pm(z),$$
$$l_+r_+ = l_-r_- = 1, \qquad l_+r_- = l_-r_+ = 0.$$

Then we can set $\tilde{P}_\pm(z) = r_\pm(z)l_\pm(z)$. The result is

$$\tilde{\lambda}_\pm(z) = \frac{\mu z}{2} \mp \rho(z),$$

(5.9)
$$\tilde{P}_\pm(z) = \begin{pmatrix} \frac{1}{2} \pm \frac{\mu z}{4\rho(z)} & \mp\frac{1}{2\rho(z)} \\ \pm\frac{p'}{2\rho(z)} & \frac{1}{2} \mp \frac{\mu z}{4\rho(z)} \end{pmatrix},$$

$$\rho(z) = \sqrt{-p' + \left(\frac{\mu z}{2}\right)^2}.$$

From (5.7) and (5.8) it is clear that

$$E(z)\tilde{P}_\pm(z) = \tilde{\lambda}_\pm(z)\tilde{P}_\pm(z).$$

Thus the solution to (5.5) is

$$(5.10) \qquad \hat{G}(\xi, t) = e^{i\xi\tilde{\lambda}_-(i\xi)t}\tilde{P}_-(i\xi) + e^{i\xi\tilde{\lambda}_+(i\xi)t}\tilde{P}_+(i\xi) = e^{i\xi E(i\xi)t}.$$

Perform the inverse Fourier transform

$$F^{-1}(\hat{g}) = \frac{1}{2\pi} \text{ pr. v.} \int_{-\infty}^{\infty} \hat{g}(\xi, t)e^{ix\xi}\, d\xi$$

to $\hat{G}$. We obtain

$$(5.11) \qquad G(x, t) = \frac{1}{2\pi} \int_{-\infty}^{\infty} \left[e^{i\xi\tilde{\lambda}_-(i\xi)t}\tilde{P}_-(i\xi) + e^{i\xi\tilde{\lambda}_+(i\xi)t}\tilde{P}_+(i\xi) \right] e^{ix\xi}\, d\xi.$$

From (5.9), we have

$$i\xi\tilde{\lambda}_{\pm}(i\xi) = -\frac{\mu}{2}\xi^2 \mp i\xi\sqrt{-p' - \left(\frac{\mu\xi}{2}\right)^2}.$$

Since $\mu > 0$ and $p' < 0$, we have

Lemma 5.1. *For real $\xi \neq 0$, $\text{Re}\{i\xi\tilde{\lambda}_{\pm}(i\xi)\} < 0$.*

Again from (5.9) we notice that $\tilde{\lambda}_{\pm}(z)$ and $\tilde{P}_{\pm}(z)$ are branches of algebraic functions of a complex variable z, having common branch points $z_{\pm} = \pm\frac{2}{\mu}\sqrt{-p'}i$ of order one. At each branch point $\tilde{\lambda}_{\pm}(z)$ are continuous while $\tilde{P}_{\pm}(z)$ have a pole there. We also notice that when z describes a small circle around z_+ or z_-, $\{\tilde{\lambda}_+(z), \tilde{\lambda}_-(z)\}$ and $\{\tilde{P}_+(z), \tilde{P}_-(z)\}$ both will undergo the transposition of themselves after analytic continuation. This implies that $\hat{G}(\xi, t)$ in (5.10) is single-valued in ξ. It is easy to show that $\hat{G}(\xi, t)$ is bounded near the isolated singularities $i\xi = z_{\pm}$:

$$|\hat{G}(\xi, t)| \leq e^{|\xi||E(i\xi)|t} \leq e^{|\xi|(|A|+|\xi||B|)t}$$
$$\rightarrow e^{\frac{2}{\mu}\sqrt{-p'}(|A|+\frac{2}{\mu}\sqrt{-p'}|B|)t} \qquad \text{as } i\xi \rightarrow z_{\pm}.$$

Hence we have proved the following lemma:

Lemma 5.2. *$\hat{G}(\xi, t)$ in (5.10) is an entire function of ξ.*

we now approximate system (5.3) by

$$(5.12) \qquad\qquad\qquad u_t + Au_x = B^*u_{xx},$$

where

$$(5.13) \qquad B^* = \begin{pmatrix} \frac{\mu}{2} & 0 \\ 0 & \frac{\mu}{2} \end{pmatrix}.$$

System (5.12) is a diagonalizable system, and its Green's function $G^*(x,t)$ can be found by diagonalization. Here, however, we go through the analysis by Fourier transform so as to compare with the estimates for G. From (5.12), $\hat{G}^*$ satisfies

$$(5.14) \qquad \begin{cases} \hat{G}_t^* = i\xi E^*(i\xi)\hat{G}^* \\ \hat{G}^*(\xi,0) = I, \end{cases}$$

where

$$E^*(z) = -A + zB^*.$$

With (5.2) and (5.13), we have the spectral representation

$$E^*(z) = \lambda_-^*(z)P_-^* + \lambda_+^*(z)P_+^*,$$

where the eigenvalues $\lambda_\pm^*$ and eigenprojections $P_\pm^*$ of E^* are

$$(5.15) \qquad \lambda_\pm^*(z) = \frac{\mu z}{2} \mp \sqrt{-p'},$$

$$P_\pm^* = \begin{pmatrix} \frac{1}{2} & \mp\frac{1}{2\sqrt{-p'}} \\ \mp\frac{\sqrt{-p'}}{2} & \frac{1}{2} \end{pmatrix}.$$

Similar to (5.10), the solution to (5.14) is

$$(5.16) \qquad \hat{G}^*(\xi,t) = e^{i\xi\lambda_-^*(i\xi)t}P_-^* + e^{i\xi\lambda_+^*(i\xi)t}P_+^*.$$

The inverse Fourier transform gives us

(5.17)

$$G^*(x,t) = \frac{1}{\sqrt{\mu t}}e^{-\frac{(x+t\sqrt{-p'})^2}{2\mu t}} \begin{pmatrix} \frac{1}{2} & \frac{1}{2\sqrt{-p'}} \\ \frac{\sqrt{-p'}}{2} & \frac{1}{2} \end{pmatrix} + \frac{1}{\sqrt{\mu t}}e^{-\frac{(x-t\sqrt{-p'})^2}{2\mu t}} \begin{pmatrix} \frac{1}{2} & -\frac{1}{2\sqrt{-p'}} \\ -\frac{\sqrt{-p'}}{2} & \frac{1}{2} \end{pmatrix}.$$

The Green's function G is basically determined by the behavior of $\tilde{\lambda}_\pm(z)$ and $\tilde{P}_\pm(z)$ when z is small and when z is large.

Lemma 5.3. *The eigenvalues $\tilde{\lambda}_\pm(z)$ and eigenprojections $\tilde{P}_\pm(z)$ of $E(z)$, given by (5.9), are holomorphic at the origin, with the property*

$$
\begin{aligned}
\tilde{\lambda}_\pm(0) + \tilde{\lambda}'_\pm(0)z &= \lambda^*_\pm(z),\\
\tilde{P}_\pm(0) &= P^*_\pm,
\end{aligned}
\tag{5.18}
$$

where $\lambda^*_\pm(z)$ and $P^*_\pm$ are eigenvalues and eigenprojections of $E^*(z)$, given by (5.15). Moreover, $\tilde{\lambda}_\pm(0)$ and $\tilde{\lambda}'_\pm(0)$ are real with $\tilde{\lambda}'_\pm(0) > 0$.

The proof of Lemma 5.3 is a simple Taylor expansion of (5.9).

Lemma 5.4. *For large z, $e^{z\tilde{\lambda}_-(z)t}\tilde{P}_-(z)$ and $e^{z\tilde{\lambda}_+(z)t}\tilde{P}_+(z)$, or the other way around, have the expressions*

$$
O(1)e^{(c_{-1}z^2+c_0z+O(1))t}
\tag{5.19a}
$$

and

$$
e^{\frac{p't}{\mu}}\begin{pmatrix}1 & 0\\0 & 0\end{pmatrix} + e^{-\alpha t}\left[z^{-1}C_M + O(z^{-2})(1 + te^{O(tz^{-1})})\right]
\tag{5.19b}
$$

respectively, where c_{-1} and α are positive constants, c_0 is a real constant, and C_M a constant matrix.

Proof. For large z, $\rho(z)$ in (5.9) has the expression

$$
\rho(z) = \pm\frac{\mu z}{2}\sqrt{1 + \frac{-p'}{(\mu z/2)^2}} = \pm\frac{\mu z}{2} \pm \frac{-p'}{\mu z} + O(z^{-3}),
\tag{5.20}
$$

where the sign on the right-hand side depends on how we define the square root of a complex number. Suppose that we have the "+" sign. Then by (5.9) and (5.20),

$$
\tilde{\lambda}_-(z) = \mu z + O(z^{-1}), \qquad \tilde{\lambda}_+(z) = \frac{p'}{\mu z} + O(z^{-3}),
$$

$$
\begin{aligned}
\tilde{P}_\pm(z) &= \begin{pmatrix}\frac{1}{2}\pm\frac{1}{2(1+O(z^{-2}))} & \mp\frac{1}{\mu z(1+O(z^{-2}))}\\ \pm\frac{p'}{\mu z(1+O(z^{-2}))} & \frac{1}{2}\mp\frac{1}{2(1+O(z^{-2}))}\end{pmatrix}\\
&= \begin{pmatrix}\frac{1}{2}\pm\frac{1}{2}(1+O(z^{-2})) & \mp\frac{1}{\mu z}(1+O(z^{-2}))\\ \pm\frac{p'}{\mu z}(1+O(z^{-2})) & \frac{1}{2}\mp\frac{1}{2}(1+O(z^{-2}))\end{pmatrix}.
\end{aligned}
\tag{5.21}
$$

Thus

$$\tilde{P}_-(z) = O(1),$$

(5.22)

$$\tilde{P}_+(z) = \begin{pmatrix} 1 & 0 \\ 0 & 0 \end{pmatrix} + z^{-1} \begin{pmatrix} 0 & -\frac{1}{\mu} \\ \frac{p'}{\mu} & 0 \end{pmatrix} + O(z^{-2}).$$

From (5.21) and (5.22) clearly $e^{z\tilde{\lambda}_-(z)t}\tilde{P}_-(z)$ has the first expression in (5.19), where $c_{-1} = \mu > 0$, $c_0 = 0$, and

(5.23) $\quad e^{z\tilde{\lambda}_+(z)t}\tilde{P}_+(z) = e^{\frac{p't}{\mu}} \begin{pmatrix} 1 & 0 \\ 0 & 0 \end{pmatrix} + e^{\frac{p't}{\mu}} \left(e^{O(z^{-2}t)} - 1\right)\tilde{P}_+(z) + e^{\frac{p't}{\mu}}\left(z^{-1}C_M + O(z^{-2})\right),$

where C_M is the second matrix in (5.22). Notice that $|e^z - 1| \le |z|e^{|z|}$. The right-hand side of (5.23) gives the second expression in (5.19) with $\alpha = -p'/\mu > 0$.

If we have the "$-$" sign in (5.20), interchanging the subscripts "$+$" and "$-$" for $\tilde{\lambda}$ and $\tilde{P}$ gives us the desired result. $\qquad\square$

Equation (5.18) suggests that G^* is the leading term in G, while (5.19) suggests that G contains a δ-function which decays exponentially in t. Set

(5.24) $$R(x,t) = G(x,t) - G^*(x,t) - e^{\frac{p't}{\mu}}\delta(x)\begin{pmatrix} 1 & 0 \\ 0 & 0 \end{pmatrix}.$$

Lemma 5.5. *For all $-\infty < x < \infty$, $t > 0$, we have*

(5.25) $\quad |R(x,t)| \le C(t+1)^{-1}\left[e^{-\frac{(x+\tilde{\lambda}_-(0)t)^2}{Ct}} + e^{-\frac{(x+\tilde{\lambda}_+(0)t)^2}{Ct}}\right] + Ct^{-\frac{1}{2}}e^{-t/C} + Ce^{-t/C}|x|.$

Proof. From (5.24), (5.11) and (5.16),

$$R(x,t) = \frac{1}{2\pi}\int_{-\infty}^{\infty}\left\{e^{i\xi\tilde{\lambda}_-(i\xi)t}\tilde{P}_-(i\xi) + e^{i\xi\tilde{\lambda}_+(i\xi)t}\tilde{P}_+(i\xi) - e^{i\xi\lambda_-^*(i\xi)t}P_-^*\right.$$

$$\left. - e^{i\xi\lambda_+^*(i\xi)t}P_+^* - e^{\frac{p't}{\mu}}\begin{pmatrix} 1 & 0 \\ 0 & 0 \end{pmatrix}\right\}e^{ix\xi}\,d\xi$$

$$= R_1 + R_2 + R_3,$$

where for $\varepsilon > 0$ small and $N > 0$ large,

(5.26)

$$R_1 = \frac{1}{2\pi} \int_{-\varepsilon}^{\varepsilon} \left\{ e^{i\xi\tilde{\lambda}_-(i\xi)t} \tilde{P}_-(i\xi) + e^{i\xi\tilde{\lambda}_+(i\xi)t} \tilde{P}_+(i\xi) - e^{i\xi\lambda_-^*(i\xi)t} P_-^* - e^{i\xi\lambda_+^*(i\xi)t} P_+^* \right\} e^{ix\xi}\, d\xi,$$

$$R_2 = \frac{1}{2\pi} \left(\int_{-\infty}^{-N} + \int_{N}^{\infty} \right) \left\{ e^{i\xi\tilde{\lambda}_-(i\xi)t} \tilde{P}_-(i\xi) + e^{i\xi\tilde{\lambda}_+(i\xi)t} \tilde{P}_+(i\xi) - e^{\frac{p't}{\mu}} \begin{pmatrix} 1 & 0 \\ 0 & 0 \end{pmatrix} \right\} e^{ix\xi}\, d\xi,$$

$$R_3 = -\frac{1}{2\pi} \left(\int_{-\infty}^{-\varepsilon} + \int_{\varepsilon}^{\infty} \right) \left\{ e^{i\xi\lambda_-^*(i\xi)t} P_-^* + e^{i\xi\lambda_+^*(i\xi)t} P_+^* \right\} e^{ix\xi}\, d\xi$$

$$+ \frac{1}{2\pi} \left(\int_{-N}^{-\varepsilon} + \int_{\varepsilon}^{N} \right) \left\{ e^{i\xi\tilde{\lambda}_-(i\xi)t} \tilde{P}_-(i\xi) + e^{i\xi\tilde{\lambda}_+(i\xi)t} \tilde{P}_+(i\xi) \right\} e^{ix\xi}\, d\xi$$

$$- \frac{1}{2\pi} \int_{-N}^{N} e^{\frac{p't}{\mu}} \begin{pmatrix} 1 & 0 \\ 0 & 0 \end{pmatrix} e^{ix\xi}\, d\xi.$$

To estimate R_1, we consider each mode separately:

(5.27a)
$$R_1 = R_{1,-} + R_{1,+},$$

where

(5.27b)
$$R_{1,\mp} = \frac{1}{2\pi} \int_{-\varepsilon}^{\varepsilon} \left\{ e^{i\xi\tilde{\lambda}_\mp(i\xi)t} \tilde{P}_\mp(i\xi) - e^{i\xi\lambda_\mp^*(i\xi)t} P_\mp^* \right\} e^{ix\xi}\, d\xi.$$

Consider $R_{1,-}$. From (5.18),

(5.28)
$$R_{1,-} = \frac{1}{2\pi} \int_{-\varepsilon}^{\varepsilon} e^{i\xi\tilde{\lambda}_-(0)t - \xi^2 \tilde{\lambda}_-'(0)t} \left[e^{O(\xi^3 t)} \tilde{P}_-(i\xi) - P_-^* \right] e^{ix\xi}\, d\xi$$

$$= \frac{1}{2\pi} \int_{-\varepsilon}^{\varepsilon} e^{i\xi\tilde{\lambda}_-(0)t - \xi^2 \tilde{\lambda}_-'(0)t + ix\xi} \left[O(\xi^3 t) e^{O(\xi^3 t)} + O(\xi) \right] d\xi.$$

Since the integrand is holomorphic at the origin by Lemma 5.3, we move the path of integration to

(5.29)
$$\sigma_- = \sigma(-\varepsilon, \varepsilon, c),$$

where

$$\sigma(a, b, c) = \{ \xi \mid \operatorname{Re}\xi = a, \ \operatorname{Im}\xi \text{ is from } 0 \text{ to } c \}$$

$$\cap \{ \xi \mid \operatorname{Im}\xi = c, \ \operatorname{Re}\xi \text{ is from } a \text{ to } b \}$$

$$\cap \{ \xi \mid \operatorname{Re}\xi = b, \ \operatorname{Im}\xi \text{ is from } c \text{ to } 0 \}$$

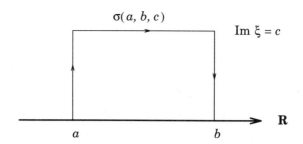

$$\sigma(a, b, c)$$

$$\text{Im } \xi = c$$

$$\mathbf{R}$$

$$a \qquad\qquad b$$

FIGURE 5.1

as shown in Figure 5.1, and c will be given later. Let $\xi = \zeta + i\eta$, ζ and η are real. Then

(5.30) $\operatorname{Re}\{i\xi\tilde{\lambda}_-(0)t - \xi^2\tilde{\lambda}'_-(0)t + ix\xi\} = -\eta(x + \tilde{\lambda}_-(0)t) - \zeta^2\tilde{\lambda}'_-(0)t + \eta^2\tilde{\lambda}'_-(0)t.$

Here we have noticed that $\tilde{\lambda}_-(0)$ and $\tilde{\lambda}'_-(0)$ are real by Lemma 5.3. Also notice that $\tilde{\lambda}'_-(0) > 0$.

For $\frac{|x+\tilde{\lambda}_-(0)t|}{2\tilde{\lambda}'_-(0)t} \leq \frac{1}{2}\varepsilon$, let $c = \frac{x+\tilde{\lambda}_-(0)t}{2\tilde{\lambda}'_-(0)t}$ in (5.29). Then by (5.28) and (5.30) we have

$$
\begin{aligned}
|R_{1,-}| &\leq \frac{1}{2\pi}\int_{-\varepsilon}^{\varepsilon} e^{-\frac{(x+\tilde{\lambda}_-(0)t)^2}{4\tilde{\lambda}'_-(0)t} - \zeta^2\tilde{\lambda}'_-(0)t}\left[O\left(|\zeta|^3 t + \frac{|x+\tilde{\lambda}_-(0)t|^3}{t^2}\right)\right.\\
&\qquad \left.\cdot e^{O\left(|\zeta|^3 t + \frac{|x+\tilde{\lambda}_-(0)t|^3}{t^2}\right)} + O\left(|\zeta| + \frac{|x+\tilde{\lambda}_-(0)t|}{t}\right)\right]d\zeta
\end{aligned}
$$

(5.31) $$+ \frac{1}{\pi}\int_0^{\frac{|x+\tilde{\lambda}_-(0)t|}{2\tilde{\lambda}'_-(0)t}} e^{-\eta|x+\tilde{\lambda}_-(0)t| - \varepsilon^2\tilde{\lambda}'_-(0)t + \eta^2\tilde{\lambda}'_-(0)t}\left[O(\varepsilon^3 t)e^{O(\varepsilon^3 t)} + O(\varepsilon)\right]d\eta$$

$$
\begin{aligned}
&\leq C\int_{-\varepsilon}^{\varepsilon} e^{-\frac{(x+\tilde{\lambda}_-(0)t)^2}{8\tilde{\lambda}'_-(0)t} - \frac{1}{2}\zeta^2\tilde{\lambda}'_-(0)t}\left[|\zeta| + \frac{|x+\tilde{\lambda}_-(0)t|}{t}\right]d\zeta\\
&\quad + Ce^{-\frac{3}{4}\varepsilon^2\tilde{\lambda}'_-(0)t}\left[\varepsilon^2 t e^{C\varepsilon^3 t} + 1\right]\varepsilon^2\\
&\leq C(t+1)^{-1}e^{-\frac{(x+\tilde{\lambda}_-(0)t)^2}{Ct}} + Ce^{-t/C}.
\end{aligned}
$$

For $\frac{|x+\tilde{\lambda}_-(0)t|}{2\tilde{\lambda}'_-(0)t} \geq \frac{1}{2}\varepsilon$, let $c = \frac{\varepsilon}{2}\operatorname{sign}(x + \tilde{\lambda}_-(0)t)$ in (5.29). Again by (5.28) and (5.30),

$$
\begin{aligned}
|R_{1,-}| &\leq C\int_{-\varepsilon}^{\varepsilon} e^{-\frac{\varepsilon}{2}|x+\tilde{\lambda}_-(0)t| - \zeta^2\tilde{\lambda}'_-(0)t + \frac{1}{4}\varepsilon^2\tilde{\lambda}'_-(0)t}\left[\varepsilon^3 t e^{C\varepsilon^3 t} + \varepsilon\right]d\zeta\\
&\quad + C\int_0^{\varepsilon/2} e^{-\eta|x+\tilde{\lambda}_-(0)t| - \varepsilon^2\tilde{\lambda}'_-(0)t + \eta^2\tilde{\lambda}'_-(0)t}\left[\varepsilon^3 t e^{C\varepsilon^3 t} + \varepsilon\right]d\eta\\
&\leq C\int_{-\varepsilon}^{\varepsilon} e^{-\frac{1}{4}\varepsilon^2\tilde{\lambda}'_-(0)t - \zeta^2\tilde{\lambda}'_-(0)t}\left[\varepsilon^3 t e^{C\varepsilon^3 t} + \varepsilon\right]d\zeta + Ce^{-t/C}\\
&\leq Ce^{-t/C}.
\end{aligned}
$$

Thus we have proved that for all $-\infty < x < \infty$, $t > 0$,

$$(5.32) \qquad |R_{1,-}| \leq C(t+1)^{-1}e^{-\frac{(x+\tilde{\lambda}_-(0)t)^2}{Ct}} + Ce^{-t/C}.$$

Similarly,

$$(5.33) \qquad |R_{1,+}| \leq C(t+1)^{-1}e^{-\frac{(x+\tilde{\lambda}_+(0)t)^2}{Ct}} + Ce^{-t/C}.$$

Inequalities (5.32) and (5.33), together with (5.27a), imply that $|R_1|$ is bounded by the right-hand side of (5.25).

Next we consider R_2. From (5.26) and Lemma 5.4,

$$|R_2| = \frac{1}{2\pi}\left|\left(\int_{-\infty}^{-N} + \int_N^\infty\right)\{O(1)e^{(-c_{-1}\xi^2 + ic_0\xi + O(1))t}\right.$$
$$\left. + e^{-\alpha t}\left[\frac{1}{i\xi}C_M + O(\xi^{-2})(1 + te^{O(t\xi^{-1})})\right]\}e^{ix\xi}\,d\xi\right|,$$

where c_{-1} and α are positive constants, c_0 is a real constant, and C_M a constant matrix. Thus

$$|R_2| \leq C\left(\int_{-\infty}^{-N} + \int_N^\infty\right)e^{-\xi^2 t/C}\,d\xi + Ce^{-\alpha t}\left|\left(\int_{-\infty}^{-N} + \int_N^\infty\right)e^{ix\xi}\frac{1}{i\xi}\,d\xi\right|$$
$$+ Ce^{-t/C}\left(\int_{-\infty}^{-N} + \int_N^\infty\right)\xi^{-2}\,d\xi$$
$$\leq Ct^{-\frac{1}{2}}e^{-t/C} + Ce^{-\alpha t}\left|2\int_N^\infty \frac{\sin x\xi}{\xi}\,d\xi\right|$$
$$\leq Ct^{-\frac{1}{2}}e^{-t/C} + Ce^{-t/C}|x|,$$

which is bounded by the right-hand side of (5.25).

To estimate R_3, notice that by (5.18) $\mathrm{Re}\{i\xi\lambda_\pm^*(i\xi)\} = -\tilde{\lambda}_\pm'(0)\xi^2$ with $\tilde{\lambda}_\pm'(0) > 0$ for real ξ. Also notice that $\mathrm{Re}\{i\xi\tilde{\lambda}_\pm(i\xi)\} \leq -1/C$ for $-N \leq \xi \leq -\varepsilon$ and $\varepsilon \leq \xi \leq N$ by Lemma 5.1 and by the fact that $\tilde{\lambda}_\pm$ are continuous functions. However, as we have pointed out before $\tilde{P}_\pm(i\xi)$ have a pole at the algebraic singularities $\xi_\pm = \frac{1}{i}z_\pm = \pm\frac{2}{\mu}\sqrt{-p'}$. Let $I_\pm$ denote the small intervals $[\xi_\pm - \varepsilon, \xi_\pm + \varepsilon]$. From (5.26),

$$(5.34) \quad |R_3| \leq C\left(\int_{-\infty}^{-\varepsilon} + \int_\varepsilon^\infty\right)\{e^{-\tilde{\lambda}_-'(0)\xi^2 t} + e^{-\tilde{\lambda}_+'(0)\xi^2 t}\}\,d\xi + Ce^{-t/C}$$
$$+ C\left|\left(\int_{I_-} + \int_{I_+}\right)\{e^{i\xi\tilde{\lambda}_-(i\xi)t}\tilde{P}_-(i\xi) + e^{i\xi\tilde{\lambda}_+(i\xi)t}\tilde{P}_+(i\xi)\}e^{ix\xi}\,d\xi\right|.$$

By (5.10) and Lemma 5.2, the integrand of the last integral in (5.34) is analytic. Hence we can move the path of integration from $I_\pm$ to

$$\sigma_\pm = \sigma(\xi_\pm - \varepsilon, \xi_\pm + \varepsilon, \varepsilon\,\text{sign}(x)),$$

see Figure 5.1. The last term in (5.34) is then bounded by

$$C\left(\int_{\sigma_-} + \int_{\sigma_+}\right)\left\{e^{\text{Re}\{i\xi\tilde\lambda_-(i\xi)\}t}|\tilde P_-(i\xi)| + e^{\text{Re}\{i\xi\tilde\lambda_+(i\xi)\}t}|\tilde P_+(i\xi)|\right\}e^{-x\eta}\,d|\xi|.$$

Take Puiseux expansion for $\tilde\lambda_\pm(i\xi)$ at ξ_- and notice that $\tilde\lambda_\pm$ are continuous there. The integral on σ_- is bounded by

$$C\int_{\sigma_-}\left\{e^{\text{Re}\{i\xi_-\tilde\lambda_-(i\xi_-)\}t + O(\varepsilon^{1/2})t}|\tilde P_-(i\xi)| + e^{\text{Re}\{i\xi_-\tilde\lambda_+(i\xi_-)\}t + O(\varepsilon^{1/2})t}|\tilde P_+(i\xi)|\right\}d|\xi|$$
$$\leq Ce^{-t/C}.$$

Similarly the intergral on σ_+ is also bounded by $Ce^{-t/C}$. It is clear that the first term in (5.34) is bounded by $Ct^{-1/2}e^{-t/C}$. Thus $|R_3|$ is bounded by the right-hand side of (5.25). $\qquad\square$

Lemma 5.6. *Let $K > 0$ be large. For $-\infty < x < \infty$, $t > 0$, if $|x|/t \geq K$, then*

$$(5.35) \qquad |R(x,t)| \leq Ct^{-\frac{1}{2}}\left(e^{-\frac{x^2}{Ct}} + e^{-\frac{\left(x+t\sqrt{-p'}\right)^2}{Ct}} + e^{-\frac{\left(x-t\sqrt{-p'}\right)^2}{Ct}}\right).$$

Proof. From (5.24),

$$(5.36) \qquad R(x,t) = R_4 - G^*(x,t),$$

where

$$R_4 = G(x,t) - e^{\frac{p't}{\mu}}\delta(x)\begin{pmatrix} 1 & 0 \\ 0 & 0 \end{pmatrix} = \frac{1}{2\pi}\int_{-\infty}^{\infty}\left\{\hat G(\xi,t) - e^{\frac{p't}{\mu}}\begin{pmatrix} 1 & 0 \\ 0 & 0 \end{pmatrix}\right\}e^{ix\xi}\,d\xi.$$

The integrand on the right-hand side is an entire function by Lemma 5.2. Hence we can move the path of integration. Set $\xi = \zeta + i\eta$ with real ζ and η. let $\nu > 0$ be a constant such that $\nu c_{-1} \leq \frac{1}{2}$, where c_{-1} is defined in Lemma 5.4. Let

$$\sigma = \sigma(-N, N, \frac{\nu x}{t}),$$

see Figure 5.1. For $|x|/t \geq K$, K large, from (5.10) and Lemma 5.4 we have

$$(5.37) \quad |R_4| = \frac{1}{2\pi} \lim_{N \to \infty} \int_\sigma \Big\{ O(1)e^{(-c_{-1}\xi^2 + ic_0\xi + O(1))t}$$
$$+ e^{-\alpha t}\Big[\frac{1}{i\xi}C_M + O(\xi^{-2})(1 + te^{O(t\xi^{-1})})\Big]\Big\}e^{ix\xi}\,d\xi,$$

where c_{-1}, $\alpha > 0$, c_0 is real, and C_M an constant matrix. The integrand on the line segments $\zeta = \pm N$ is bounded by

$$C\Big\{e^{-c_{-1}N^2t + c_{-1}\nu^2x^2/t + O(1)(|x|+t)} + e^{-\alpha t}\Big[\frac{1}{N} + \frac{1}{N^{-2}}(1 + te^{Ct/N})\Big]\Big\},$$

which goes to zero as $N \to \infty$ for fixed x and t. Therefore, in (5.37) the integral on the line segments $\zeta = \pm N$ goes to zero as $N \to \infty$. Equation (5.37) becomes

$$R_4 = \frac{1}{2\pi} \int_{-\infty}^{\infty} \Big\{ O(1)e^{(-c_{-1}(\zeta + i\nu x/t)^2 + ic_0(\zeta + i\nu x/t) + O(1))t}$$
$$+ e^{-\alpha t}\Big[\frac{1}{i\zeta - \nu x/t}C_M + O\Big(\frac{1}{\zeta^2 + (\nu x/t)^2}\Big)(1 + te^{O(t/K)})\Big]\Big\}e^{ix(\zeta + i\nu x/t)}\,d\zeta.$$

Notice that $\nu c_{-1} \leq \frac{1}{2}$ and K is large. We have

$$|R_4| \leq C \int_{-\infty}^{\infty} e^{-c_{-1}\zeta^2 t + c_{-1}\nu^2 \frac{x^2}{t} - c_0\nu x + O(1)t - \nu \frac{x^2}{t}}\,d\xi + Ce^{-\alpha t - \nu \frac{x^2}{t}}\Big|\int_{-\infty}^{\infty} \frac{e^{ix\zeta}}{i\zeta - \nu x/t}\,d\zeta\Big|$$
$$+ C \int_{-\infty}^{\infty} \frac{e^{-\alpha t}}{\zeta^2 + (\nu K)^2}(1 + te^{Ct/K})e^{-\nu \frac{x^2}{t}}\,d\zeta$$
$$\leq Ce^{-\frac{\nu x^2}{2t} + C|x|}\int_{-\infty}^{\infty} e^{-c_{-1}\zeta^2 t}\,d\zeta + Ce^{-\alpha t - \frac{\nu x^2}{t}}\Big|\int_{-\infty}^{\infty} e^{ix\zeta}\Big[\frac{1}{i\zeta}$$
$$- \frac{\nu x}{t}\Big(\frac{1}{\zeta^2 + (\nu x/t)^2} + \frac{1}{i\zeta}\frac{\nu x/t}{\zeta^2 + (\nu x/t)^2}\Big)\Big]\,d\zeta\Big| + Ce^{-t/C - \frac{\nu x^2}{t}}$$
$$\leq Ct^{-\frac{1}{2}}e^{-\frac{\nu x^2}{2t} + \frac{Cx^2}{Kt}} + Ce^{-\alpha t - \frac{\nu x^2}{t}}\Big(1 + \nu \frac{|x|}{t} + \Big(\frac{\nu x}{t}\Big)^2|x|\Big) + Ce^{-t/C - \frac{x^2}{Ct}}$$
$$\leq Ct^{-\frac{1}{2}}e^{-\frac{x^2}{Ct}}.$$

Thus $|R_4|$ is bounded by the right-hand side of (5.35). From (5.17) it is clear that $G^*(x,t)$ is also bounded by the right-hand side of (5.35). Equation (5.36) then implies (5.35). $\square$

we now have the main theorem of this section.

Theorem 5.7. [Ze] *For all* $-\infty < x < \infty$, $t \geq 0$, *the Green's function* G *of the p-system (5.1) has the property*

$$(5.38) \quad G(x,t) = G^*(x,t) + O(1)(t+1)^{-\frac{1}{2}} t^{-\frac{1}{2}} \left(e^{-\frac{\left(x+t\sqrt{-p'}\right)^2}{Ct}} + e^{-\frac{\left(x-t\sqrt{-p'}\right)^2}{Ct}} \right)$$

$$+ e^{\frac{p'}{\mu}t} \delta(x) \begin{pmatrix} 1 & 0 \\ 0 & 0 \end{pmatrix},$$

where G^* *is the Green's function of the uniform parabolic system (5.12), given by (5.17), and* δ *is the Dirac* δ-*function.*

Proof. Fix a large K in Lemma 5.6. If $|x|/t \geq K$, by (5.35) we have for a constant $C_1 > 0$,

$$(5.39)$$

$$|R(x,t)| \leq C t^{-\frac{1}{2}} \left(e^{-\frac{x^2}{C_1 t}} + e^{-\frac{\left(x+t\sqrt{-p'}\right)^2}{C_1 t}} + e^{-\frac{\left(x-t\sqrt{-p'}\right)^2}{C_1 t}} \right)$$

$$\leq C t^{-\frac{1}{2}} \left[e^{-\frac{\left(x+t\sqrt{-p'}\right)^2}{2C_1\left(1+\sqrt{-p'}/K\right)^2 t} - \frac{x^2}{2C_1 t}} + \left(e^{-\frac{\left(x+t\sqrt{-p'}\right)^2}{2C_1 t}} + e^{-\frac{\left(x-t\sqrt{-p'}\right)^2}{2C_1 t}} \right) e^{-\frac{\left(K-\sqrt{-p'}\right)^2 t}{2C_1}} \right]$$

$$\leq C(t+1)^{-\frac{1}{2}} t^{-\frac{1}{2}} \left(e^{-\frac{\left(x+t\sqrt{-p'}\right)^2}{Ct}} + e^{-\frac{\left(x-t\sqrt{-p'}\right)^2}{Ct}} \right).$$

If $|x|/t \leq K$, by (5.25) we have

$$(5.40)$$

$$|R(x,t)| \leq C(t+1)^{-1} \left(e^{-\frac{\left(x+t\sqrt{-p'}\right)^2}{C_1 t}} + e^{-\frac{\left(x-t\sqrt{-p'}\right)^2}{C_1 t}} \right) + C t^{-\frac{1}{2}} e^{-t/C_1} + C e^{-t/C_1} |x|$$

$$\leq C(t+1)^{-\frac{1}{2}} t^{-\frac{1}{2}} \left(e^{-\frac{\left(x+t\sqrt{-p'}\right)^2}{Ct}} + e^{-\frac{\left(x-t\sqrt{-p'}\right)^2}{Ct}} \right)$$

since $\tilde{\lambda}_{\mp}(0) = \pm\sqrt{-p'}$ by (5.9) and

$$e^{-t/C_1} \leq e^{-\frac{t}{2C_1} - \frac{\left(x+t\sqrt{-p'}\right)^2}{2C_1\left(K+\sqrt{-p'}\right)^2 t}}.$$

Equation (5.24), together with (5.39) and (5.40), gives us (5.38). $\qquad \square$

We have similar result for the derivatives of G.

Theorem 5.8. *For all* $-\infty < x < \infty$, $t \geq 0$, *the Green's function* G *of the p-system (5.1) has the property*

$$(5.41) \quad \frac{\partial^l}{\partial x^l} G(x,t) = \frac{\partial^l}{\partial x^l} G^*(x,t) + O(1)(t+1)^{-\frac{1}{2}} t^{-\frac{l+1}{2}} \left(e^{-\frac{\left(x+t\sqrt{-p'}\right)^2}{Ct}} + e^{-\frac{\left(x-t\sqrt{-p'}\right)^2}{Ct}} \right)$$

$$+ e^{\frac{p'}{\mu}t} \sum_{j=0}^{l} \delta^{(l-j)}(x) Q_j,$$

where $l \geq 0$ is an integer, G^* given by (5.17), and Q_j, $j = 0, \ldots, l$, are 2×2 polynomial matrices in t with degrees not more than j. Especially,

$$Q_0 = \begin{pmatrix} 1 & 0 \\ 0 & 0 \end{pmatrix}.$$

Proof. We have from (5.10) and properties of the Fourier transform,

$$(5.42) \qquad \frac{\partial^l}{\partial x^l} G(x,t) = F^{-1}\{(i\xi)^l e^{i\xi\tilde{\lambda}_-(i\xi)t}\tilde{P}_-(i\xi) + (i\xi)^l e^{i\xi\tilde{\lambda}_+(i\xi)t}\tilde{P}_+(i\xi)\}.$$

Similar to Lemma 5.4, one of $z^l e^{z\tilde{\lambda}_-(z)t}\tilde{P}_-(z)$ and $z^l e^{z\tilde{\lambda}_+(z)t}\tilde{P}_+(z)$ has the expression

$$O(1)z^l e^{(c_{-1}z^2 + c_0 z + O(1))t}$$

for large z, while the other has the expression

$$e^{\frac{p't}{\mu}}\sum_{j=0}^{l} z^{l-j}Q_j + e^{-\alpha t}\big[z^{-1}Q_{l+1} + O(z^{-2})(1 + \cdots + t^{l+1}) + O(z^{-2}t^{l+2})e^{O(tz^{-1})}\big],$$

where c_{-1} and α are positive constants, c_0 is a real constant, and Q_j, $j = 0, \ldots, l+1$, are 2×2 polynomial matrices in t with degrees not more than j. Especially, Q_0 is given explicitly as in the theorem. Set

$$(5.43) \qquad R^{(l)}(x,t) = \frac{\partial^l}{\partial x^l} G(x,t) - \frac{\partial^l}{\partial x^l} G^*(x,t) - e^{\frac{p't}{\mu}}\sum_{j=0}^{l}\delta^{(l-j)}(x)Q_j.$$

As in the proof of Lemma 5.5, we write $R^{(l)}$ by (5.10) and (5.16) as

$$R^{(l)} = \frac{1}{2\pi}\int_{-\infty}^{\infty}\Big\{(i\xi)^l\big[e^{i\xi\tilde{\lambda}_-(i\xi)t}\tilde{P}_-(i\xi) + e^{i\xi\tilde{\lambda}_+(i\xi)t}\tilde{P}_+(i\xi) - e^{i\xi\lambda_-^*(i\xi)t}P_-^*$$

$$(5.44) \qquad\qquad - e^{i\xi\lambda_+^*(i\xi)t}P_+^*\big] - e^{\frac{p't}{\mu}}\sum_{j=0}^{l}(i\xi)^{l-j}Q_j\Big\}e^{ix\xi}\,d\xi$$

$$= R_1^{(l)} + R_2^{(l)} + R_3^{(l)},$$

where for $\varepsilon > 0$ small and $N > 0$ large,

$$R_1^{(l)} = \frac{1}{2\pi}\int_{-\varepsilon}^{\varepsilon}(i\xi)^l\{e^{i\xi\tilde{\lambda}_-(i\xi)t}\tilde{P}_-(i\xi) + e^{i\xi\tilde{\lambda}_+(i\xi)t}\tilde{P}_+(i\xi) - e^{i\xi\lambda_-^*(i\xi)t}P_-^*$$

$$- e^{i\xi\lambda_+^*(i\xi)t}P_+^*\}e^{ix\xi}\,d\xi,$$

$$R_2^{(l)} = \frac{1}{2\pi}\left(\int_{-\infty}^{-N} + \int_{N}^{\infty}\right)\left\{(i\xi)^l e^{i\xi\tilde\lambda_-(i\xi)t}\tilde P_-(i\xi) + (i\xi)^l e^{i\xi\tilde\lambda_+(i\xi)t}\tilde P_+(i\xi)\right.$$

$$\left. - e^{\frac{p't}{\mu}}\sum_{j=0}^{l}(i\xi)^{l-j}Q_j\right\}e^{ix\xi}\,d\xi,$$

$$R_3^{(l)} = -\frac{1}{2\pi}\left(\int_{-\infty}^{-\varepsilon} + \int_{\varepsilon}^{\infty}\right)(i\xi)^l\left\{e^{i\xi\lambda_-^*(i\xi)t}P_-^* + e^{i\xi\lambda_+^*(i\xi)t}P_+^*\right\}e^{ix\xi}\,d\xi$$

$$+ \frac{1}{2\pi}\left(\int_{-N}^{-\varepsilon} + \int_{\varepsilon}^{N}\right)(i\xi)^l\left\{e^{i\xi\tilde\lambda_-(i\xi)t}\tilde P_-(i\xi) + e^{i\xi\tilde\lambda_+(i\xi)t}\tilde P_+(i\xi)\right\}e^{ix\xi}\,d\xi$$

$$- \frac{1}{2\pi}\int_{-N}^{N} e^{\frac{p't}{\mu}}\sum_{j=0}^{l}(i\xi)^{l-j}Q_j e^{ix\xi}\,d\xi.$$

The estimating for $R_2^{(l)}$ and $R_3^{(l)}$ is almost the same as for R_2 and R_3 in the proof of Lemma 5.5, where we notice that

$$\int_{\varepsilon}^{\infty}\xi^l e^{-\xi^2 t/C}\,d\xi = O(1)t^{-\frac{l+1}{2}}e^{-t/C}.$$

The result is

(5.45) $$|R_2^{(l)}| + |R_3^{(l)}| \le Ct^{-\frac{l+1}{2}}e^{-t/C} + Ce^{-t/C}|x|.$$

To estimate $R_1^{(l)}$, we have $R_1^{(l)} = R_{1,-}^{(l)} + R_{1,+}^{(l)}$, where

$$R_{1,\mp}^{(l)} = \frac{1}{2\pi}\int_{-\varepsilon}^{\varepsilon}(i\xi)^l\left\{e^{i\xi\tilde\lambda_\mp(i\xi)t}\tilde P_\mp(i\xi) - e^{i\xi\lambda_\mp^*(i\xi)t}P_\mp^*\right\}e^{ix\xi}\,d\xi.$$

Compared with the definition of $R_{1,\mp}$ in (5.27b), we see that the integrand for $R_{1,\mp}^{(l)}$ has an additional factor $(i\xi)^l$. Following the same procedure for $R_{1,-}$, we have for $\frac{|x+\tilde\lambda_-(0)t|}{2\tilde\lambda_-'(0)t} \le \frac{1}{2}\varepsilon$,

$$|R_{1,-}^{(l)}| \le C\int_{-\varepsilon}^{\varepsilon}e^{-\frac{(x+\tilde\lambda_-(0)t)^2}{8\tilde\lambda_-'(0)t} - \frac{1}{2}\zeta^2\tilde\lambda_-'(0)t}\left[|\zeta|^{l+1} + \frac{|x+\tilde\lambda_-(0)t|^{l+1}}{t^{l+1}}\right]d\zeta + Ce^{-t/C}$$

$$\le C(t+1)^{-\frac{l+2}{2}}e^{-\frac{(x+\tilde\lambda_-(0)t)^2}{Ct}} + Ce^{-t/C},$$

cf. (5.31). For $\frac{|x+\tilde\lambda_-(0)t|}{2\tilde\lambda_-'(0)t} \ge \frac{1}{2}\varepsilon$, $R_{1,-}^{(l)}$ has the same bound $Ce^{-t/C}$ as $R_{1,-}$. Similarly we can estimate $R_{1,+}^{(l)}$. Together with (5.44) and (5.45), we thus have

(5.46) $$|R^{(l)}| \le C(t+1)^{-\frac{l+2}{2}}\left[e^{-\frac{(x+\tilde\lambda_-(0)t)^2}{Ct}} + e^{-\frac{(x+\tilde\lambda_+(0)t)^2}{Ct}}\right] + Ct^{-\frac{l+1}{2}}e^{-t/C} + Ce^{-t/C}|x|,$$

TAI-PING LIU AND YANNI ZENG

which is an analogue of (5.25). Similar to (5.35), we can also prove that for $|x|/t \geq K$, K large,

$$(5.47) \qquad |R^{(l)}| \leq Ct^{-\frac{l+1}{2}} \left(e^{-\frac{x^2}{Ct}} + e^{-\frac{\left(x+t\sqrt{-p'}\right)^2}{Ct}} + e^{-\frac{\left(x-t\sqrt{-p'}\right)^2}{Ct}} \right).$$

Equation (5.43), together with (5.46) and (5.47), implies (5.41).   □

6. Green's Functions for $n \times n$ Systems with Applications

We now extend the result of last section to general $n \times n$ hyperbolic-parabolic linear systems. At the end of this section we give applications to the Navier-Stokes equations and the equations of magnetohydrodynamics.

6.1. Green's Functions.

We consider the general system

$$(6.1) \qquad u_t + Au_x = Bu_{xx}, \qquad -\infty < x < \infty, \quad t > 0,$$

where $u = u(x,t)$ is an n-vector, while A and B are $n \times n$ constant matrices. Equation (6.1) can be viewed as the linearization of (2.1). Our basic assumptions for (6.1) are the following linear version of Assumptions 2.1 and 2.3.

Assumption 6.1. There exists a symmetric positive definite matrix A_0, such that $A_0 A$ is symmetric, and $A_0 B$ is symmetric semi-positive definite.

Assumption 6.2. Any eigenvector of A is not in the null space of B.

The purpose of this subsection is to obtain an estimate similar to (5.38) for the Green's function $G(x,t)$ of system (6.1), which is the solution matrix to (6.1) satisfying the initial condition $G(x,0) = \delta(x)I$.

As we have pointed out in Section 2, Assumption 6.1 implies that all the eigenvalues of A are real, and that A has a complete set of eigenvectors. We use the same notations here as in Section 2. Denote all the distinct eigenvalues of A as $\lambda_1, \lambda_2, \ldots, \lambda_s$, with multiplicity $m_1, m_2, \ldots, m_s, m_1+m_2+\cdots+m_s = n$. Denote the left and right eigenvectors associated with λ_i as l_{ij} and r_{ij} respectively, $j = 1, \ldots, m_i$, satisfying

$$(6.2) \qquad \begin{aligned} &Ar_{ij} = \lambda_i r_{ij}, \\ &l_{ij}A = \lambda_i l_{ij}, \\ &l_{ij}r_{i'j'} = \delta_{ii'}\delta_{jj'}, \\ &i,i' = 1,\ldots,s, \quad j = 1,\ldots,m_i, \quad j' = 1,\ldots,m_{i'}. \end{aligned}$$

We also use the notations

$$l_i = \begin{pmatrix} l_{i1} \\ \vdots \\ l_{im_i} \end{pmatrix}, \qquad r_i = (r_{i1}, \ldots, r_{im_i}),$$

(6.3) $$P_i = r_i l_i, \qquad i = 1, \ldots, s,$$

$$L = \begin{pmatrix} l_1 \\ \vdots \\ l_s \end{pmatrix}, \qquad R = (r_1, \ldots, r_s).$$

Here P_i is the eigenprojection of A associated with λ_i, which is unique. Clearly $LR = I$ by (6.2).

We now approximate (6.1) by

(6.4) $$u_t + Au_x = B^* u_{xx},$$

where

(6.5)
$$B^* = RDL, \qquad D = \mathrm{diag}(D_1, \ldots, D_s),$$
$$D_i = l_i B r_i \in \mathbb{R}^{m_i \times m_i}, \quad i = 1, \ldots, s.$$

By Lemma 2.1, we see that under Assumptions 6.1 and 6.2 each D_i is similar to a diagonal matrix with positive diagonal elements. Hence there exist $T_i \in \mathbb{R}^{m_i \times m_i}$, $i = 1, \ldots, s$, such that

$$T_i^{-1} D_i T_i = \begin{pmatrix} \mu_{i1} \\ & \ddots \\ & & \mu_{im_i} \end{pmatrix} \equiv \mu_i, \qquad i = 1, \ldots, s,$$

where $\mu_{ij} > 0$, $j = 1, \ldots, m_i$, are eigenvalues of D_i. Denote the identity in $\mathbb{R}^{m_i \times m_i}$ as I_i, and set

$$T = \mathrm{diag}(T_1, \ldots, T_s),$$
$$M = \mathrm{diag}(\mu_1, \ldots, \mu_s),$$
$$\Lambda = \mathrm{diag}(\lambda_1 I_1, \ldots, \lambda_s I_s).$$

Multiply (6.4) by $T^{-1}L$ on the left. We thus diagonalize it as

$$(T^{-1}Lu)_t + \Lambda(T^{-1}Lu)_x = M(T^{-1}Lu)_{xx}.$$

The Green's function of this system is straightforward, so is that of (6.4). We notice that the row vectors of $T_i^{-1} l_i$ are left eigenvectors of A, and the column vectors of $r_i T_i$ are right eigenvectors, associated with λ_i. Denote them as l_{ij}^* and r_{ij}^* respectively, $i = 1, \ldots, s$, $j = 1, \ldots, m_i$. Clearly l_{ij}^* and r_{ij}^* satisfy (6.2). We thus have

Lemma 6.1. *The green's function of (6.4) is*

$$(6.6) \qquad G^*(x,t) = \sum_{i=1}^{s} \sum_{j=1}^{m_i} \frac{1}{\sqrt{4\pi\mu_{ij}t}} e^{-\frac{(x-\lambda_i t)^2}{4\mu_{ij}t}} r^*_{ij} l^*_{ij},$$

where for $i = 1, \dots, s$, $j = 1, \dots, m_i$, λ_i, r^*_{ij} and l^*_{ij} are the eigenvalues, right eigenvectors and left eigenvectors of A satisfying (6.2) and

$$\begin{pmatrix} l^*_{i1} \\ \vdots \\ l^*_{im_i} \end{pmatrix} B(r^*_{i1}, \dots, r^*_{im_i}) = \begin{pmatrix} \mu_{i1} & & \\ & \ddots & \\ & & \mu_{im_i} \end{pmatrix}, \qquad i = 1, \dots, s,$$

with the $\mu_{ij} > 0$.

Notice that Assumption 6.1 also guarantees that all the eigenvalues of B are nonnegative, and that B has a complete set of eigenvectors. We are now ready to give the main theorem of this section.

Theorem 6.2. *Suppose that Assumptions 6.1 and 6.2 are satisfied, and that B has the zero eigenvalue of multiplicity m, $0 \le m < n$. Let l_{0j} and r_{0j} be the left and right eigenvectors, respectively, of B associated with the zero eigenvalue, satisfying $l_{0j} r_{0k} = \delta_{jk}$, j, $k = 1$, $\dots$, m. Then for $-\infty < x < \infty$, $t \ge 0$, the Green's function G of system (6.1) has the property*

$$(6.7) \qquad G(x,t) = G^*(x,t) + O(1)(t+1)^{-\frac{1}{2}} t^{-\frac{1}{2}} \sum_{j=1}^{s} e^{-\frac{(x-\lambda_j t)^2}{Ct}} + \sum_{k=1}^{m'} e^{-\alpha_k t}\delta(x - \beta_k t)\mathcal{P}_k,$$

where G^* given by (6.6) is the Green's function of (6.4); λ_j, $j = 1, \dots, s$, are the distinct eigenvalues of A; $C > 0$ is a constant; $m' \le m$; for $k = 1, \dots, m'$, $\alpha_k > 0$ and β_k are constants, $\mathcal{P}_k$ are constant projections, satisfying

$$\mathcal{P}_k \mathcal{P}_{k'} = \delta_{kk'} \mathcal{P}_k, \qquad k, k' = 1, \dots, m',$$

$$(6.8) \qquad \sum_{k=1}^{m'} \mathcal{P}_k = \sum_{j=1}^{m} r_{0j} l_{0j} \equiv Q_0,$$

$$\sum_{k=1}^{m'} \beta_k \mathcal{P}_k = Q_0 A Q_0,$$

and δ is the Dirac δ-function.

Remark 6.3. If B is nonsingular, $m' = 0$ and the second summation in (6.7) disappears. We may say that the δ-functions in (6.7) are due to the hyperbolic ingredient of the system.

Remark 6.4. Suppose that B is singular. From (6.7) we see that $dx/dt = \beta_k$, $k = 1, \ldots ,$ m', are the directions in which an initial jump will propagate. From (6.8) the β_k are the eigenvalues of $Q_0 A Q_0$ when restricted to the range of Q_0. We can also compute m', α_k and $\mathcal{P}_k$, $k = 1, \ldots , m'$, explicitly using a reduction process. In fact, let $\nu_1, \ldots , \nu_\tau$ be all the nonzero (and hence positive) eigenvalues of B with corresponding eigenprojections $Q_1, \ldots , Q_\tau$ (cf. P_i in (6.3)). Set

$$(6.9) \qquad S = \sum_{j=1}^{\tau} \frac{1}{\nu_j} Q_j,$$

which is the value at zero of the reduced resolvent of B with respect to the zero eigenvalue. Associated with each distinct eigenvalue β_k, $Q_0 A Q_0$ has the eigenprojection $\mathcal{Q}_k$. Here if β_k happens to be zero, $\mathcal{Q}_k$ is taken as the subprojection $\mathcal{Q}_k Q_0$. Corresponding to this β_k there may be several α_k, which are all the nonzero distinct eigenvalues of $\mathcal{Q}_k A S A \mathcal{Q}_k$. Each α_k then associates with an eigenprojection of $\mathcal{Q}_k A S A \mathcal{Q}_k$, which is the corresponding $\mathcal{P}_k$. At last m' is the total number of the α_k so constructed for all the distinct β_k. In the special case that all the β_k are simple, we have

$$(6.10) \qquad m' = m, \quad \mathcal{P}_k \text{ determined by (6.8)}, \quad \alpha_k = \operatorname{tr}(A S A \mathcal{P}_k), \quad k = 1, \ldots, m.$$

Remark 6.5. From Remark 6.4 we see that if B is replaced by μB in (6.1) with $\mu > 0$, then $\alpha_k = O(\mu^{-1})$ while β_k and $\mathcal{P}_k$ are independent of μ, $k = 1, \ldots , m'$. This can also be seen from $G(x, t; \mu) = \frac{1}{\mu} G(x/\mu, t/\mu; 1)$, which comes from (6.13) below. Therefore, in (6.7) the δ-functions disappear when $\mu \to 0$. We may say that the δ-functions are due to the parabolic nature in addition to the hyperbolic ingredient (Remark 6.3) of the system.

The proof of Theorem 6.2 consists of a series of lemmas. First we notice that under Assumption 6.1 we may decompose A_0 as $A_0 = (A_0^{\frac{1}{2}})^2$, where $A_0^{\frac{1}{2}}$ is a symmetric and positive definite matrix. Introduce a new variable $w = A_0^{\frac{1}{2}} u$. Then w satisfies a symmetric system

$$w_t + A_0^{\frac{1}{2}} A (A_0^{\frac{1}{2}})^{-1} w_x = A_0^{\frac{1}{2}} B (A_0^{\frac{1}{2}})^{-1} w_{xx}$$

since $A_0 A$, $A_0 B$ and $(A_0^{\frac{1}{2}})^{-1}$ are all symmetric. Therefore it is sufficient to prove Theorem 6.2 for symmetric systems. For the rest of this subsection we assume that A and B are symmetric and $A_0 = I$.

Perform the Fourier transform with respect to x for G. From (6.1) we have

$$\begin{cases} \hat{G}_t = i\xi E(i\xi)\hat{G} \\ \hat{G}(\xi,0) = I, \end{cases}$$

where

(6.11) $$E(z) = -A + zB.$$

The solution is

(6.12) $$\hat{G}(\xi,t) = e^{i\xi E(i\xi)t}.$$

Hence

(6.13) $$G(x,t) = F^{-1}\{e^{i\xi E(i\xi)t}\}.$$

From the case of the p-system we see that G is determined by the behavior of the eigenvalues and eigenprojections of E. We now establish a series of lemmas analogous to Lemmas 5.1–5.4. The following lemma is due to Shizuta and Kawashima.

Lemma 6.6 [SK]. *Under Assumption 6.1, Assumption 6.2 is equivalent to any of the following statements:*

(i) There exists a $K \in \mathbb{R}^{n \times n}$, such that KA_0 is skew symmetric and $\frac{1}{2}(KA_0A + A_0AK^t) + A_0B$ is positive definite.

(ii) Let $\tilde{\lambda}(z)$ be any eigenvalue of $E(z)$ defined by (6.11). Then $\mathrm{Re}\{i\xi\tilde{\lambda}(i\xi)\} < 0$ for real $\xi \neq 0$.

(iii) There exists a constant $c > 0$, such that $\mathrm{Re}\{i\xi\tilde{\lambda}(i\xi)\} \leq -c\xi^2/(1 + \xi^2)$ for real ξ, where $\tilde{\lambda}(z)$ is any eigenvalue of $E(z)$.

Proof. we have assumed that A and B are symmetric and $A_0 = I$. For the general case, we can prove the lemma by making use of the decomposition $A_0 = (A_0^{\frac{1}{2}})^2$.

We first prove that Assumption 6.2 implies (i). Define an inner product $\{\cdot, \cdot\}$ in $\mathbb{R}^{n \times n}$ as

$$\{X, Y\} = \mathrm{tr}(XY^t)$$

for any $X, Y \in \mathbb{R}^{n \times n}$. Define a linear transform $\Gamma_A \colon \mathbb{R}^{n \times n} \to \mathbb{R}^{n \times n}$ as

$$\Gamma_A(X) = AX - XA, \qquad \forall X \in \mathbb{R}^{n \times n}.$$

Denote the null space and range of Γ_A as S_1 and S_2 respectively. We have

$$(6.14) \qquad \qquad \dim S_1 + \dim S_2 = \dim \mathbb{R}^{n \times n}.$$

For any $X \in S_1$, $AY - YA \in S_2$,

$$\{X, AY - YA\} = \mathrm{tr}(XY^t A - XAY^t) = \mathrm{tr}(XY^t A - AXY^t)$$
$$= \mathrm{tr}(XY^t A) - \mathrm{tr}(AXY^t) = 0,$$

where we have used the facts that A is symmetric, $AX = XA$ since $X \in S_1$, and $\mathrm{tr}(ST) = \mathrm{tr}(TS)$. Thus $S_1 \perp S_2$. Together with (6.14), we conclude that $S_2 = S_1^\perp$.

By (6.2) and (6.3), we have the spectral representation

$$(6.15) \qquad \qquad A = \sum_{j=1}^{s} \lambda_j P_j,$$

It is easy to see that each P_j is symmetric since A is symmetric. If $Y \in \mathbb{R}^{n \times n}$ commutes with A, Y must commute with P_j. In fact, multiply $YA = AY$ from the left by P_j and from the right by P_k, use (6.15) and notice that $P_j P_k = \delta_{jk} P_j$. We arrive at $\lambda_k P_j Y P_k = \lambda_j P_j Y P_k$, or $P_j Y P_k = 0$ for $j \neq k$. This implies that $P_j Y = P_j Y \sum_{k=1}^{s} P_k = P_j Y P_j$. Similarly we obtain $Y P_j = P_j Y P_j$ and therefore $Y P_j = P_j Y$.

Define a linear transform $\Pi_A \colon \mathbb{R}^{n \times n} \to \mathbb{R}^{n \times n}$ as

$$(6.16) \qquad \qquad \Pi_A(X) = \sum_{j=1}^{s} P_j X P_j, \qquad \forall X \in \mathbb{R}^{n \times n}.$$

For any $X \in \mathbb{R}^{n \times n}$, $Y \in S_1$,

$$\{X - \Pi_A(X), Y\} = \mathrm{tr}(XY^t - \Pi_A(X)Y^t) = \mathrm{tr}(XY^t) - \mathrm{tr}(\sum_{j=1}^{s} P_j X P_j Y^t)$$
$$= \mathrm{tr}(XY^t) - \mathrm{tr}(\sum_{j=1}^{s} Y P_j X^t P_j) = \mathrm{tr}(XY^t) - \mathrm{tr}(\sum_{j=1}^{s} P_j Y X^t P_j)$$
$$= \mathrm{tr}(XY^t) - \mathrm{tr}(\sum_{j=1}^{s} Y X^t P_j^2) = \mathrm{tr}(XY^t) - \mathrm{tr}(YX^t) = 0.$$

Hence $X - \Pi_A(X) \in S_1^\perp = S_2$. Especially, there exists a $K \in \mathbb{R}^{n \times n}$, such that

$$B - \Pi_A(B) = AK - KA.$$

Set $K_1 = \frac{1}{2}(K + K^t)$, $K_2 = \frac{1}{2}(K - K^t)$. Then

$$B - \Pi_A(B) = A(K_1 + K_2) - (K_1 + K_2)A.$$

The left-hand side is symmetric, so is the right-hand side, which implies that $AK_1 - K_1 A = 0$. Thus

$$K_2 A - AK_2 + B = \Pi_A(B).$$

From (6.16), using the notations in (6.3) and (6.5), we have $\Pi_A(B) = RDL = B^*$. Assumptions 6.1 and 6.2 imply that all the eigenvalues of B^* are positive, as we have pointed out before, hence $\Pi_A(B)$ is symmetric and positive definite. Set $K = 2K_2$ and obtain (i).

We next prove that (i) implies (iii). For real ξ, let $\tilde{\lambda}(i\xi)$ be any eigenvalue of $E(i\xi)$ with eigenvector $r(i\xi)$:

$$(6.17) \qquad\qquad\qquad (-A + i\xi B)r = \tilde{\lambda} r.$$

Multiplying (6.17) from the left by $i\xi \bar{r}^t$ and taking real part, we have

$$(6.18) \qquad\qquad\qquad \mathrm{Re}\{i\xi\tilde{\lambda}\}|r|^2 + \xi^2 \bar{r}^t Br = 0,$$

where we have used the fact that $\bar{r}^t Sr$ is real if $S \in \mathbb{R}^{n \times n}$ is symmetric. Multiplying (6.17) from the left by $\xi^2 \bar{r}^t K$, K given in (i), and taking real part, we have

$$(6.19) \quad -i\xi\,\mathrm{Re}\{i\xi\tilde{\lambda}\}\bar{r}^t Kr + \frac{1}{2}\xi^2 \bar{r}^t (KA + AK^t)r = \mathrm{Re}(i\xi^3 \bar{r}^t KBr)$$

$$\leq |\xi|^3 |r| \|K\| |Br| \leq \frac{\varepsilon}{2}\xi^2 |r|^2 + \frac{1}{2\varepsilon}\xi^4 \|K\|^2 |Br|^2,$$

where $\varepsilon > 0$ and we have used the fact that $\bar{r}^t Sr$ is pure imaginary is $S \in \mathbb{R}^{n \times n}$ is skew symmetric. Multiplying (6.18) by $(1 + \xi^2)$ and (6.19) by $\alpha > 0$, addition of the results gives us

$$(6.20) \quad \mathrm{Re}\{i\xi\tilde{\lambda}\}(1 + \xi^2)\left(|r|^2 - \frac{i\xi\alpha}{1+\xi^2}\bar{r}^t Kr\right) + \xi^2\left[\bar{r}^t Br + \bar{r}^t \frac{\alpha}{2}(KA + AK^t)r\right] + \xi^4 \bar{r}^t Br$$

$$\leq \frac{\varepsilon\alpha}{2}\xi^2 |r|^2 + \frac{\alpha}{2\varepsilon}\xi^4 \|K\|^2 |Br|^2.$$

Clearly from (i) and Assumption 6.1 that B is semi-positive definite we have

$$\bar{r}^t B r + \frac{1}{2}\bar{r}^t(KA + AK^t)r \geq c_1|r|^2,$$

$$|Br|^2 \leq c_2\bar{r}^t Br, \qquad \left|-\frac{i\xi\alpha}{1+\xi^2}\bar{r}^t Kr\right| \leq c_3\alpha|r|^2,$$

where $c_j > 0$, $j = 1, 2, 3$, are constants. Applying these inequalities to (6.20) and choosing $\varepsilon = c_1$, $\alpha < \min\{1, \frac{2c_1}{c_2|K|^2}, \frac{1}{c_3}\}$, we have

$$\operatorname{Re}\{i\xi\tilde{\lambda}\}(1+\xi^2)\left(|r|^2 - \frac{i\xi\alpha}{1+\xi^2}\bar{r}^t Kr\right) + \frac{c_1\alpha}{2}\xi^2|r|^2 \leq 0,$$

$$\operatorname{Re}\{i\xi\tilde{\lambda}\} \leq -\frac{c_1\alpha/2}{1+c_3\alpha}\frac{\xi^2}{1+\xi^2}.$$

Hence (iii) is proved with $c = \frac{1}{2}c_1\alpha/(1 + c_3\alpha)$.

Obviously (iii) implies (ii). At last we prove (ii) implies Assumption 6.2. If Assumption 6.2 were false, there would be an $r \in \mathbb{R}^n$, such that $r \neq 0$, $Br = 0$, and $Ar = \lambda r$ for some real λ. Thus

$$(-A + zB)r = -\lambda r,$$

or there would be a $\tilde{\lambda}(z) = -\lambda$. Then $\operatorname{Re}\{i\xi\tilde{\lambda}(i\xi)\} = 0$ for any real ξ. This contradicts (ii). $\square$

We now state some basic facts from complex analysis and perturbation theory. The reader is referred to Kato's book [Kt], Chapter 2, and the references therein.

Consider $E(z)$ given by (6.11) as a matrix-valued function of the complex variable z. The number s' of the distinct eigenvalues of $E(z)$ is constant if z is not one of the exceptional points, of which there are only a finite number in the whole complex plane. In each simply connected domain containing no exceptional points, the eigenvalues of $E(z)$ can be expressed as s' holomorphic functions

$$\tilde{\lambda}_1(z), \tilde{\lambda}_2(z), \ldots, \tilde{\lambda}_{s'}(z),$$

which have constant multiplicities $\tilde{m}_1, \tilde{m}_2, \ldots, \tilde{m}_{s'}$ respectively, $\sum\limits_{j=1}^{s'} \tilde{m}_j = n$. The $\tilde{\lambda}_j(z)$, $1 \leq j \leq s'$, are branches of one or several algebraic functions, which are everywhere continuous since the coefficient of the highest power in the characteristic equation is 1.

These algebraic functions will also be denoted by $\tilde{\lambda}_j(z)$. An exceptional point z_0 is either a branch point of some of the $\tilde{\lambda}_j(z)$ or a regular point for all of them; in the latter case the values of some of the different $\tilde{\lambda}_j(z)$ coincide at $z = z_0$. Hence there is always splitting of eigenvalues at (and only at) an exceptional point.

In general, $E(z)$ of the form (6.11) is not diagonalizable, and has the spectral representation for any z which is not an exceptional point:

$$(6.21) \qquad E(z) = \sum_{j=1}^{s'} \tilde{\lambda}_j(z)\tilde{P}_j(z) + \sum_{j=1}^{s'} \tilde{D}_j(z),$$

where

$$(6.22) \qquad \tilde{\lambda}_j \neq \tilde{\lambda}_k \quad \text{for } j \neq k, \qquad \tilde{P}_j\tilde{P}_k = \delta_{jk}\tilde{P}_j, \qquad \sum_{j=1}^{s'} \tilde{P}_j = I,$$

$$\tilde{D}_j^{\tilde{m}_j} = 0, \qquad \tilde{P}_j\tilde{D}_j = \tilde{D}_j\tilde{P}_j = \tilde{D}_j, \qquad j, k = 1, \ldots, s'.$$

The spectral representation of a matrix is unique. Here $\tilde{P}_j$ and $\tilde{D}_j$, called the eigenprojection and eigennilpotent for $\tilde{\lambda}_j$ respectively, must be given by

$$\tilde{P}_j(z) = -\frac{1}{2\pi i}\int_{\Gamma_j(z)} (E(z) - \xi I)^{-1} d\xi,$$
$$\tilde{D}_j(z) = (E(z) - \tilde{\lambda}_j(z)I)\tilde{P}_j(z),$$

where $\Gamma_j(z)$ is a positive-oriented circle enclosing $\tilde{\lambda}_j(z)$ but excluding all other $\tilde{\lambda}_k(z)$.

The eigenprojections $\tilde{P}_j(z)$ and the eigennilpotents $\tilde{D}_j(z)$ for $\tilde{\lambda}_j(z)$ are also holomorphic in each simply connected domain containing no exceptional points, being branches of one or several algebraic functions (again denoted by $\tilde{P}_j(z)$ and $\tilde{D}_j(z)$).

The algebraic functions $\tilde{P}_j(z)$ and $\tilde{\lambda}_j(z)$ have common branch points of the same order. Let z_0 be an exceptional point. When z describes a small circle around z_0, $\{\tilde{\lambda}_1(z), \tilde{\lambda}_2(z), \ldots, \tilde{\lambda}_{s'}(z)\}$ and $\{\tilde{P}_1(z), \tilde{P}_2(z), \ldots, \tilde{P}_{s'}(z)\}$ will undergo the same permutation among themselves after analytic continuation. We emphasize here that z_0 may or may not be a branch point, $\tilde{\lambda}_j(z)$ is always continuous there, while $\tilde{P}_j(z)$ in general has a pole at z_0 even when z_0 is not a branch point. If z_0 indeed is a branch point, we have the following lemma due to Butler.

Lemma 6.7. *([Bu] or [Kt]) If $z = z_0$ is a branch point of $\tilde{\lambda}_j(z)$ (and therefore also of $\tilde{P}_j(z)$) of order $p - 1 \geq 1$, then $\tilde{P}_j(z)$ has a pole there; that is, the Laurent expansion of $\tilde{P}_j(z)$ in powers of $(z - z_0)^{1/p}$ necessarily contains negative powers.*

Proof. Without loss of generality we assume that $z_0 = 0$. When z is subjected to a revolution around $z = 0$, $\tilde{P}_j(z)$ is changed into, say, $\tilde{P}_{j+1}(z)$. Suppose that the lemma were not true. Let

$$\tilde{P}_k(z) = \tilde{P}_k^{(0)} + z^{1/p}\tilde{P}_k^{(1)} + \cdots, \qquad k = j, j+1$$

be the Laurent expansions of $\tilde{P}_j$ and $\tilde{P}_{j+1}$. We must have $\tilde{P}_{j+1}^{(l)} = e^{i2\pi l/p}\tilde{P}_j^{(l)}$, $l = 0, 1,$ $\ldots$, since $\tilde{P}_j(z)$ becomes $\tilde{P}_{j+1}(z)$ after the revolution of z. Especially, $\tilde{P}_{j+1}^{(0)} = \tilde{P}_j^{(0)}$. On the other hand, the relation $\tilde{P}_j(z)\tilde{P}_{j+1}(z) = 0$ in (6.22) for $z \to 0$ implies $\tilde{P}_j^{(0)}\tilde{P}_{j+1}^{(0)} = 0$, and the idempotent character of $\tilde{P}_j(z)$ implies $\left(\tilde{P}_j^{(0)}\right)^2 = \tilde{P}_j^{(0)}$. Hence $\tilde{P}_j^{(0)} = \left(\tilde{P}_j^{(0)}\right)^2 = \tilde{P}_j^{(0)}\tilde{P}_{j+1}^{(0)} = 0$, and $\lim_{z \to 0}|\tilde{P}_j(z)| = 0$. We notice that any eigenprojection is not zero. In fact, the rank of $\tilde{P}_j(z)$ is $\tilde{m}_j$. Thus (6.22) implies that $|\tilde{P}_j(z)| \geq 1$. Therefor we reach a contradiction, and Lemma 6.7 is proved. $\qquad\square$

We now explore the $\tilde{\lambda}_j(z)$, $\tilde{P}_j(z)$ and $\tilde{D}_j(z)$ under Assumptions 6.1 and 6.2.

Lemma 6.8. *Assume Assumption 6.1. For $j = 1, \ldots, s'$, $\tilde{D}_j(z) \equiv 0$, while $\tilde{\lambda}_j(z)$ and $\tilde{P}_j(z)$ are holomorphic at the origin. At the infinity, the $\tilde{\lambda}_j(z)$ are single-valued, having at most a pole of order 1, while the $\tilde{P}_j(z)$ are holomorphic there.*

Proof. Under Assumption 6.1, both A and B are symmetric. Thus $E(z)$ defined by (6.11) is real symmetric for real z. We then have an orthonormal set of eigenvectors for $E(z)$. For any real z which is not an exceptional point, let the orthonormal set be $\tilde{r}_{jk}(z)$, $j = 1,$ $\ldots$, s', $k = 1, \ldots, \tilde{m}_j$, where the $\tilde{r}_{jk}$ are eigenvectors associated with $\tilde{\lambda}_j$. We set

$$(6.23) \qquad \tilde{P}_j = \sum_{k=1}^{\tilde{m}_j} \tilde{r}_{jk}\tilde{r}_{jk}^t, \qquad \tilde{D}_j = 0, \qquad j = 1, \ldots, s',$$

which clearly satisfy (6.21) and (6.22). The uniqueness of spectral representation then implies that the $\tilde{P}_j$ and $\tilde{D}_j$ in (6.23) are the ones under consideration. Notice that we only have a finite number of exceptional points. After analytic continuation, $\tilde{D}_j(z) = 0$ on the whole complex plane. Meanwhile, from (6.23) $|\tilde{P}_j(z)| = 1$ for real z which is not an exceptional point. By Lemma 6.4 the analytic functions $\tilde{\lambda}_j(z)$ and $\tilde{P}_j(z)$ can not have real branch points. Especially, they are single-valued at the origin. Therefore, the $\tilde{\lambda}_j(z)$ are holomorphic there since they are continuous, and the $\tilde{P}_j(z)$ are also holomorphic there since they are bounded. At the infinity, we write

$$E(z) = z(-\frac{1}{z}A + B) \equiv z\tilde{E}(z).$$

Then $\tilde{E}(z)$ has eigenvalues $\frac{1}{z}\tilde{\lambda}_j(z)$ and eigenprojections $\tilde{P}_j(z)$, $j = 1, \ldots, s'$, by (6.21). As we have proved, $\frac{1}{z}\tilde{\lambda}_j(z)$ and $\tilde{P}_j(z)$ are holomorphic at the infinity ($\frac{1}{z} \to 0$). Hence we have completed the proof of Lemma 6.8. □

By Lemma 6.8, the spectral representation (6.21), (6.22) becomes

(6.24)
$$E(z) = \sum_{j=1}^{s'} \tilde{\lambda}_j(z)\tilde{P}_j(z),$$

$$\tilde{P}_j\tilde{P}_k = \delta_{jk}\tilde{P}_j, \quad j,k = 1, \ldots, s', \quad \sum_{j=1}^{s'} \tilde{P}_j = I,$$

where z is not an exceptional point. Equations (6.12) and (6.24) give us

(6.25)
$$\hat{G}(\xi,t) = \sum_{j=1}^{s'} e^{i\xi\tilde{\lambda}_j(i\xi)t}\tilde{P}_j(i\xi).$$

Following the same argument for Lemma 5.2, we obtain

Lemma 6.9. $\hat{G}(\xi,t)$ in (6.25) is an entire function of ξ.

We further discuss the behavior of the $\tilde{\lambda}_j(z)$ and $\tilde{P}_j(z)$ when z is small and when z is large.

Lemma 6.10. *Under Assumption 6.1, we have the following expressions:*

(6.26)
$$-A = \sum_{j=1}^{s'} \tilde{\lambda}_j(0)\tilde{P}_j(0),$$

(6.27)
$$B^* = \sum_{j=1}^{s'} \tilde{\lambda}'_j(0)\tilde{P}_j(0),$$

(6.28)
$$P_jBP_j = \sum_k \tilde{\lambda}'_k(0)\tilde{P}_k(0), \qquad j = 1, \ldots, s,$$

where B^* is defined in (6.5), P_j in (6.3), and the summation in (6.28) is for all the k such that $\tilde{\lambda}_k(0) = -\lambda_j$, $1 \leq k \leq s'$.

Proof. By Lemma 6.8 the $\tilde{\lambda}_j(z)$ and $\tilde{P}_j(z)$ are holomorphic at the origin. As $z \to 0$ in (6.24), we obtain (6.26) by (6.11).

From (6.5) and (6.3) we have

$$(6.29) \qquad\qquad B^* = \sum_{j=1}^{s} r_j D_j l_j = \sum_{j=1}^{s} P_j B P_j.$$

From (6.2) and (6.3) we have the spectral representation

$$(6.30) \qquad\qquad A = \sum_{j=1}^{s} \lambda_j P_j.$$

Equations (6.26), (6.30) and the uniqueness of spectral representation imply that we can classify the $\tilde{\lambda}_j(z)$ and $\tilde{P}_j(z)$ as the $\tilde{\lambda}_{jk}(z)$ and $\tilde{P}_{jk}(z)$, such that

$$
-\tilde{\lambda}_{jk}(0) = \lambda_j, \qquad j = 1, \ldots, s, \quad k = 1, \ldots, n_j,
$$

$$(6.31) \qquad\qquad \sum_{k=1}^{n_j} \tilde{P}_{jk}(0) = P_j, \qquad j = 1, \ldots, s,$$

$$
n_1 + \cdots + n_s = s'.
$$

Multiplying $E(z)$ by $\sum_{k=1}^{n_j} \tilde{P}_{jk}(z)$ from the left and from the right and using (6.11) and (6.24), we have for $j = 1, \ldots, s$,

$$
\left(\sum_{k=1}^{n_j} \tilde{P}_{jk}(z) \right)(-A + zB)\left(\sum_{l=1}^{n_j} \tilde{P}_{jl}(z) \right) = \left(\sum_{k=1}^{n_j} \tilde{P}_{jk}(z) \right)\left(\sum_{l=1}^{n_j} \tilde{\lambda}_{jl}(z)\tilde{P}_{jl}(z) \right).
$$

Having the Taylor expansion for small z, comparing the coefficients for z on both sides, and using (6.31) and (6.24), we have

$$
\left(\sum_{k=1}^{n_j} \tilde{P}'_{jk}(0) \right)(-A)P_j + P_j B P_j + P_j(-A)\left(\sum_{l=1}^{n_j} \tilde{P}'_{jl}(0) \right)
$$

$$
= \left(\sum_{k=1}^{n_j} \tilde{P}'_{jk}(0) \right)(-\lambda_j P_j) + \sum_{l=1}^{n_j} \tilde{\lambda}'_{jl}(0)\tilde{P}_{jl}(0) + (-\lambda_j P_j)\left(\sum_{l=1}^{n_j} \tilde{P}'_{jl}(0) \right).
$$

Since $AP_j = P_j A = \lambda_j P_j$, we arrive at

$$
P_j B P_j = \sum_{k=1}^{n_j} \tilde{\lambda}'_{jk}(0)\tilde{P}_{jk}(0),
$$

which is (6.28). Summation for j, $1 \leq j \leq s$, together with (6.29) and (6.31), gives (6.27). $\qquad\qquad\qquad\qquad\qquad\qquad\qquad\qquad\qquad\qquad\qquad\qquad\qquad\quad\square$

Lemma 6.11. *Let $G^*(x,t)$ be the Green's function of (6.4). Under Assumptions 6.1 and 6.2, the Fourier transform $\hat{G}^*$ of G^* has the expression*

$$(6.32) \qquad \hat{G}^*(\xi,t) = \sum_{j=1}^{s'} e^{i\xi(\tilde{\lambda}_j(0)+i\xi\tilde{\lambda}_j'(0))t}\tilde{P}_j(0),$$

where the $\tilde{\lambda}_j(0)$ are real and the $\tilde{\lambda}_j'(0)$ are positive.

Proof. Clearly from (6.4) $\hat{G}^*(\xi,t) = e^{i\xi E^*(i\xi)t}$, where

$$E^*(z) = -A + zB^* = \sum_{j=1}^{s'}\big(\tilde{\lambda}_j(0) + z\tilde{\lambda}_j'(0)\big)\tilde{P}_j(0)$$

by (6.26) and (6.27). Notice that $\tilde{P}_j(0)\tilde{P}_k(0) = \delta_{jk}\tilde{P}_j(0)$ by (6.24). Expression (6.32) is then straightforward. From (6.26) and (6.27) we see that the $\tilde{\lambda}_j(0)$ and $\tilde{\lambda}_j'(0)$ are eigenvalues of $-A$ and B^* respectively. The $\tilde{\lambda}_j(0)$ are real since A is symmetric. The $\tilde{\lambda}_j'(0)$ are positive by Assumptions 6.1 and 6.2, see the explanation following (6.5). $\qquad\square$

Lemma 6.12. *Under Assumptions 6.1 and 6.2, for large z we have*

(6.33)

$$e^{z\tilde{\lambda}_j(z)t}\tilde{P}_j(z) = \begin{cases} e^{-(\beta_j z+\alpha_j)t}\mathcal{P}_j + e^{-(\beta_j z+\alpha_j)t}\left[z^{-1}C_{M,j} + O(z^{-2})(1+t+t^2 e^{O(tz^{-1})})\right], \\ \qquad\qquad 1 \le j \le m', \\ O(1)e^{(c_{j,-1}z^2 + c_{j,0}z + O(1))t}, \qquad m' < j \le s', \end{cases}$$

where $m' \le m$, m is the multiplicity of the zero eigenvalue of B; for $k = 1, \dots, m'$ and $j = m'+1, \dots, s'$, $\alpha_k, \beta_k, c_{j,-1}$ and $c_{j,0}$ are real constants with $\alpha_k, c_{j,-1} > 0$; for $k = 1, \dots, m'$, $\mathcal{P}_k$ are constant matrices while $C_{M,k}$ are polynomial matrices in t with degrees not more than 1; the β_k and $\mathcal{P}_k$ satisfy (6.8). Here we have arranged $\tilde{\lambda}_j(z)$, $j = 1, \dots, s'$, in appropriate order.

Proof. By Lemma 6.8, as $z \to \infty$,

(6.34)
$$\tilde{\lambda}_j(z) = c_{j,-1}z + c_{j,0} + c_{j,1}\frac{1}{z} + c_{j,2}\frac{1}{z^2} + \cdots,$$
$$\tilde{P}_j(z) = P_{j,0} + P_{j,1}\frac{1}{z} + \cdots, \qquad j = 1,\dots,s'.$$

By Assumption 6.1, $\tilde{\lambda}_j(z)$ and $\tilde{P}_j(z)$ are real for real z since $E(z)$ is real symmetric. Therefore, all the coefficients in (6.34) are real. Comparing the leading coefficients on both sides of (6.24), using (6.34), we have

$$B = \sum_{j=1}^{s'} c_{j,-1} P_{j,0},$$

(6.35)

$$P_{j,0} P_{k,0} = \delta_{jk} P_{j,0}, \quad j,k = 1,\ldots,s', \qquad \sum_{j=1}^{s'} P_{j,0} = I.$$

Hence $c_{j,-1}$, $j = 1, \ldots, s'$, are eigenvalues of B, which are nonnegative by Assumption 6.1. After rearranging the $\tilde{\lambda}_j(z)$, we have $c_{k,-1} = 0$, $1 \le k \le m'$, and $c_{j,-1} > 0$, $m'+1 \le j \le s'$, with $m' \le m$. Set

(6.36) $$\beta_k = -c_{k,0}, \qquad \alpha_k = -c_{k,1}, \qquad \mathcal{P}_k = P_{k,0}, \qquad k = 1, \ldots, m'.$$

We prove that $\alpha_k > 0$. In fact, for large real ξ we have from (6.34) and Lemma 6.6 (iii),

$$\operatorname{Re}\{i\xi \tilde{\lambda}_k(i\xi)\} = -\alpha_k + O\left(\frac{1}{\xi}\right) \le -\frac{c\xi^2}{1+\xi^2}, \qquad k = 1, \ldots, m',$$

where $c > 0$ is a constant. This implies that $\alpha_k > 0$.

For $m' < j \le s'$, the second expression in (6.33) is straightforward by (6.34). For $1 \le k \le m'$, by (6.34) and (6.36),

$$
\begin{aligned}
e^{z\tilde{\lambda}_k(z)t} \tilde{P}_k(z) &= e^{-(\beta_k z + \alpha_k - c_{k,2}\frac{1}{z} + O(\frac{1}{z^2}))t}\left(\mathcal{P}_k + P_{k,1}\frac{1}{z} + O(\frac{1}{z^2})\right) \\
&= e^{-(\beta_k z + \alpha_k)t}\left\{\mathcal{P}_k + \frac{1}{z}(c_{k,2}t\mathcal{P}_k + P_{k,1}) + O(\frac{1}{z^2})\left(1 + t + t^2 e^{O(tz^{-1})}\right)\right\},
\end{aligned}
$$

which gives us the first expression in (6.33) with $C_{M,k} = c_{k,2}t\mathcal{P}_k + P_{k,1}$.

Clearly the $\mathcal{P}_k$ satisfy the first two equations in (6.8) by (6.36), (6.35) and the uniqueness of spectral representation. Notice that the $\frac{1}{z}\tilde{\lambda}_j(z)$ and $\tilde{P}_j(z)$ are the eigenvalues and eigenprojections, respectively, of

(6.37) $$\frac{1}{z} E(z) = B + \frac{1}{z}(-A).$$

Regard $w = 1/z$ as a new variable and compare (6.37) with (6.11). We then apply (6.28) to (6.37) to obtain

$$Q_0(-A)Q_0 = \sum_{k=1}^{m'} c_{k,0} P_{k,0} = \sum_{k=1}^{m'} (-\beta_k)\mathcal{P}_k$$

using (6.34) and (6.36). The third equation in (6.8) is proved. □

We now have Lemmas 6.6, 6.8, 6.9, 6.11 and 6.12 analogous to Lemmas 5.1–5.4 in the last section. Set

$$(6.38) \qquad R(x,t) = G(x,t) - G^*(x,t) - \sum_{k=1}^{m'} e^{-\alpha_k t}\delta(x - \beta_k t)\mathcal{P}_k.$$

By (6.25) and (6.32),

$$R(x,t) = \frac{1}{2\pi}\int_{-\infty}^{\infty}\left\{\sum_{j=1}^{s'} e^{i\xi\tilde{\lambda}_j(i\xi)t}\tilde{P}_j(i\xi) - \sum_{j=1}^{s'} e^{i\xi(\tilde{\lambda}_j(0)+i\xi\tilde{\lambda}_j'(0))t}\tilde{P}_j(0)\right.$$
$$\left. - \sum_{k=1}^{m'} e^{-(\beta_k i\xi+\alpha_k)t}\mathcal{P}_k\right\}e^{ix\xi}\,d\xi.$$

With tiny modifications in the proofs for Lemmas 5.5 and 5.6, we obtain

Lemma 6.13. *Under Assumptions 6.1 and 6.2, for all $-\infty < x < \infty$, $t > 0$, we have*

$$(6.39)\quad |R(x,t)| \le C(t+1)^{-1}\sum_{j=1}^{s'} e^{-\frac{(x+\tilde{\lambda}_j(0)t)^2}{Ct}} + Ct^{-1/2}e^{-t/C} + Ce^{-t/C}\sum_{k=1}^{m'}|x - \beta_k t|.$$

Lemma 6.14. *Let $K > 0$ be large. Under Assumptions 6.1 and 6.2, for all $-\infty < x < \infty$, $t > 0$, if $|x|/t \ge K$, then*

$$(6.40)\qquad |R(x,t)| \le Ct^{-\frac{1}{2}}\left(e^{-\frac{x^2}{Ct}} + \sum_{j=1}^{s} e^{-\frac{(x-\lambda_j t)^2}{Ct}}\right).$$

As Theorem 5.7 follows from Lemmas 5.5 and 5.6, (6.7) follows from (6.38)–(6.40). Theorem 6.2 is proved. We further have the following analogue of Theorem 5.8, which extends Theorem 6.2 to higher derivatives.

Theorem 6.15. *Under the same assumptions and same notations as in Theorem 6.2, the Green's function G of system (6.1) has the property*

$$(6.41)\quad \frac{\partial^l}{\partial x^l}G(x,t) = \frac{\partial^l}{\partial x^l}G^*(x,t) + O(1)(t+1)^{-\frac{1}{2}}t^{-\frac{l+1}{2}}\sum_{j=1}^{s} e^{-\frac{(x-\lambda_j t)^2}{Ct}}$$
$$+ \sum_{k=1}^{m'} e^{-\alpha_k t}\sum_{j=0}^{l}\delta^{(l-j)}(x - \beta_k t)Q_{kj},$$

where $l \geq 0$ is an integer, $-\infty < x < \infty$, $t \geq 0$, and Q_{kj}, $k = 1, \ldots, m'$, $j = 0, \ldots, l$, are $n \times n$ polynomial matrices in t with degrees not more than j. Especially, $Q_{k0} = \mathcal{P}_k$, $k = 1, \ldots, m'$, with $\mathcal{P}_k$ being the same as in Theorem 6.2.

We finish this subsection with a justification of the computation of m', α_k and $\mathcal{P}_k$, which was described in Remark 6.4. The reader is referred to Kato's book [Kt]

We have taken m' as the number of $\frac{1}{z}\tilde{\lambda}_j(z)$ which splits from zero at $z = \infty$ in Lemma 6.12. Then some of the pairs (β_k, α_k) in (6.36) may be identical. Suppose that in (6.7) we sum up those terms corresponding to identical pairs and let m' be the number of the distinct ones. Now we use a new notation m^* to denote the number of $\frac{1}{z}\tilde{\lambda}_j(z)$ splitting from zero at $z = \infty$. Set

$$(6.42) \qquad P(z) = \sum_{k=1}^{m^*} \tilde{P}_k(z) = P^{(0)} + \frac{1}{z}P^{(1)} + \cdots.$$

Then

$$(6.43) \qquad P^{(0)} = Q_0, \qquad P^{(1)} = SAQ_0 + Q_0 AS,$$

where S is given by (6.9). The first equality is clear by (6.34), (6.36) and (6.8). To prove the second equality, by (6.11), (6.24) and (6.42) we have

$$(-A + zB)P(z) = \sum_{k=1}^{m^*} \tilde{\lambda}_k(z)\tilde{P}_k(z).$$

Comparing the constant terms on both sides, using (6.34) and noticing $c_{k,-1} = 0$, $k = 1, \ldots, m^*$, we arrive at

$$-AP^{(0)} + BP^{(1)} = \sum_{k=1}^{m^*} c_{k,0}P_{k,0}.$$

Clearly $SP_{k,0} = SQ_0 P_{k,0} = 0$ by (6.36), (6.8) and (6.9). Thus

$$(6.44) \qquad -SAP^{(0)} + SBP^{(1)} = 0.$$

Similarly we have

$$(6.45) \qquad -P^{(0)}AS + P^{(1)}BS = 0.$$

Addition of (6.44) and (6.45) yields

(6.46) $$SBP^{(1)} + P^{(1)}BS = SAP^{(0)} + P^{(0)}AS.$$

Notice that under the notations in Remark 6.4, we have the spectral representation $B = \sum_{j=1}^{\tau} \nu_j Q_j$ and thus $SB = BS = I - Q_0$. Also notice that $P^{(0)}P^{(1)} + P^{(1)}P^{(0)} = P^{(1)}$ by comparing the coefficients of z^{-1} in $(P(z))^2 = P(z)$. Equation (6.46) then gives us the second equality in (6.43) by using the first one.

Consider the matrix-valued function

(6.47) $$H(z^{-1}) \equiv P(z)E(z)P(z) = -H_0 + z^{-1}H_1 + O(z^{-2})$$

for large z, where by (6.42), (6.11), (6.43) and (6.9),

(6.48)
$$H_0 = Q_0 A Q_0,$$
$$H_1 = -Q_0 A P^{(1)} - P^{(1)} A Q_0 + P^{(1)} B P^{(1)} = -Q_0 A S A Q_0 - H_0 A S - S A H_0.$$

Clearly $H(z^{-1}) = \sum_{k=1}^{m^*} \tilde{\lambda}_k(z)\tilde{P}_k(z)$ by (6.42) and (6.24). Thus all the nonzero eigenvalues of $H(z^{-1})$ are

$$\tilde{\lambda}_k(z) = c_{k,0} + c_{k,1}z^{-1} + \cdots, \qquad k = 1, \ldots, m^*,$$

by (6.34) and the meaning of m^*. By (6.36) we may assume $-\beta_1 = c_{1,0} = \cdots = c_{\rho,0} \neq c_{j,0}$, $\rho < j \leq m^*$. We can apply (6.28) to $H(z^{-1})$ even it has a higher order term in (6.47). Under the notations in Remark 6.4, the result is

$$\mathcal{Q}_1 H_1 \mathcal{Q}_1 = \sum_{k=1}^{\rho} c_{k,1} P_{k,0} = \sum_k -\alpha_k \mathcal{P}_k,$$

where we have used (6.34) and (6.36), and summed up those terms with the same $c_{k,1}$ to make the α_k distinct. In case $\beta_1 = 0$, we multiply the equation by Q_0 from the left and from the right. Notice that the right-hand side is unchanged by (6.8). Then we use the same notaton $\mathcal{Q}_1$ to denote the subprojection $\mathcal{Q}_1 Q_0 = Q_0 \mathcal{Q}_1$. Thus whatever β_1 is, $\mathcal{Q}_1 = \mathcal{Q}_1 Q_0 = Q_0 \mathcal{Q}_1$. By (6.48), the above equation becomes

(6.49) $$\sum_k \alpha_k \mathcal{P}_k = -\mathcal{Q}_1 Q_0 H_1 Q_0 \mathcal{Q}_1 = \mathcal{Q}_1 A S A \mathcal{Q}_1.$$

The α_k and $\mathcal{P}_k$ associated with β_1 are so uniquely determined. In the same way we obtain all the α_k and $\mathcal{P}_k$ for the distinct β_k, and hence the total number m' of the α_k.

In the special case that all the β_k are simple, we have m distinct β_k since the range of Q_0 is m-dimensional. Thus $m' = m$, and the $\mathcal{P}_k$ are uniquely determined by (6.8). Notice that $\mathcal{Q}_k = \mathcal{P}_k$. Equation (6.49) becomes $\alpha_1 \mathcal{P}_1 = \mathcal{P}_1 A S A \mathcal{P}_1$. The rank of $\mathcal{P}_1$ is 1. Thus $\alpha_1 = \mathrm{tr}(\alpha_1 \mathcal{P}_1) = \mathrm{tr}(A S A \mathcal{P}_1)$, using the fact that $\mathrm{tr}(ST) = \mathrm{tr}(TS)$.

6.2. Applications.

We now consider two important examples in continuum mechanics, the Navier-Stokes equations and the equations of magnetohydrodynamics. These equations will be studied in more details in Section 9. Here we simply apply the result of last subsection to obtain the Green's functions of their linearizations.

First consider the linearization of the Navier-Stokes equations (9.1) around a constant state $(v^*, 0, e^*)$ with $v^*, e^* > 0$:

(6.50) $$w_t + A w_x = B w_{xx},$$

where

$$w = (v - v^*, u, e + \frac{1}{2}u^2 - e^*)^t,$$

(6.51)
$$A = \begin{pmatrix} 0 & -1 & 0 \\ p_v & 0 & p_e \\ 0 & p & 0 \end{pmatrix}, \qquad B = \begin{pmatrix} 0 & 0 & 0 \\ 0 & \frac{\mu}{v^*} & 0 \\ \frac{\kappa}{v^*}\theta_v & 0 & \frac{\kappa}{v^*}\theta_e \end{pmatrix},$$

and $p > 0$, $\theta_e > 0$, $\kappa > 0$, $\mu \geq 0$, p_v, p_e and θ_v are constants. In consistency of (9.3) and (9.5), we assume

(6.52) $\quad \bar{p}_v \equiv p_v - pp_e < 0, \qquad \tilde{p}_v \equiv p_v - \dfrac{\theta_v}{\theta_e}p_e < 0, \qquad p_e(\theta_v - p\theta_e) < 0.$

The matrix A has eigenvalues

(6.53) $$\lambda_1 = -\sqrt{-\bar{p}_v}, \qquad \lambda_2 = 0, \qquad \lambda_3 = \sqrt{-\bar{p}_v}.$$

The right eigenvectors r_i and left eigenvectors l_i, $i = 1, 2, 3$, for A are

$$r_i = (1, -\lambda_i, -p)^t, \quad i = 1, 3,$$

$$r_2 = (p_e, 0, -p_v)^t,$$

(6.54)
$$l_i = \frac{1}{2\bar{p}_v}(p_v, \lambda_i, p_e), \quad i = 1, 3,$$

$$l_2 = -\frac{p}{\bar{p}_v}(1, 0, \frac{1}{p}),$$

satisfying $l_i r_j = \delta_{ij}$, $i, j = 1, 2, 3$. Assumption 6.1 is implied by Assumption 2.1 for the nonlinear system, which will be varified in Section 9. Assumption 6.2 is satisfied by (6.51) and (6.54). Applying Theorem 6.2 we obtain the Green's function G of (6.50), (6.51),

(6.55)
$$G(x,t) = \sum_{i=1}^{3} \frac{1}{\sqrt{4\pi\mu_i t}} e^{-\frac{(x-\lambda_i t)^2}{4\mu_i t}} r_i l_i + O(1)(t+1)^{-\frac{1}{2}} t^{-\frac{1}{2}} \sum_{i=1}^{3} e^{-\frac{(x-\lambda_i t)^2}{Ct}} + H,$$

where $\mu_i = l_i B r_i$, and $H = \sum_{k=1}^{m'} e^{-\alpha_k t} \delta(x - \beta_k t) \mathcal{P}_k$. Direct computation gives

(6.56)
$$\mu_i = \frac{1}{2v^* \bar{p}_v}(\mu\bar{p}_v + \kappa p_e(\theta_v - p\theta_e)) > 0, \quad i = 1, 3,$$

$$\mu_2 = \frac{\kappa\theta_e \tilde{p}_v}{v^* \bar{p}_v} > 0.$$

To compute H, we consider two cases.

Case (i). $\mu > 0$. In this case $m' = m = 1$. The right eigenvector r_0 and left eigenvector l_0 of B associated with the zero eigenvalue are

$$r_0 = (1, 0, -\theta_v/\theta_e)^t, \qquad l_0 = (1, 0, 0),$$

satisfying $l_0 r_0 = 1$. From Theorem 6.2 and Remark 6.4,

$$\beta = l_0 A r_0 = 0, \qquad \mathcal{P} = r_0 l_0,$$

$$S = \frac{v^*}{\mu}(0, 1, 0)^t(0, 1, 0) + \frac{v^*}{\kappa\theta_e}(0, 0, 1)^t(\theta_v/\theta_e, 0, 1),$$

$$\alpha = \operatorname{tr}(ASA\mathcal{P}) = l_0 ASA r_0 = -\frac{v^*}{\mu}\tilde{p}_v.$$

Thus in (6.55),

(6.57)
$$H = e^{\frac{v^*}{\mu}\tilde{p}_v t} \delta(x) \begin{pmatrix} 1 & 0 & 0 \\ 0 & 0 & 0 \\ -\frac{\theta_v}{\theta_e} & 0 & 0 \end{pmatrix}.$$

Case (ii). $\mu = 0$. In this case $m = 2$. The right eigenvectors r_{0j} and left eigenvectors l_{0j}, $j = 1, 2$, of B associated with the zero eigenvalue are

$$r_{01} = (1, 0, -\theta_v/\theta_e)^t, \qquad\qquad r_{02} = (0, 1, 0)^t,$$

$$l_{01} = (1, 0, 0), \qquad\qquad l_{02} = (0, 1, 0),$$

satisfying $l_{0j} r_{0k} = \delta_{jk}$, $j, k = 1, 2$. The β_k are the eigenvalues of

$$(6.58) \qquad \binom{l_{01}}{l_{02}} A(r_{01}, r_{02}) = \begin{pmatrix} 0 & -1 \\ \tilde{p}_v & 0 \end{pmatrix},$$

i.e.

$$(6.59) \qquad \beta_1 = -\sqrt{-\tilde{p}_v}, \qquad \beta_2 = \sqrt{-\tilde{p}_v}.$$

Both β_1 and β_2 are simple. Thus $m' = 2$. We have from (6.58)

$$Q_0 A Q_0 = (r_{01}, r_{02}) \begin{pmatrix} 0 & -1 \\ \tilde{p}_v & 0 \end{pmatrix} \binom{l_{01}}{l_{02}} = \beta_1 \mathcal{P}_1 + \beta_2 \mathcal{P}_2,$$

where

$$\mathcal{P}_k = (r_{01}, r_{02}) \binom{1}{-\beta_k} (\frac{1}{2}, -\frac{1}{2\beta_k}) \binom{l_{01}}{l_{02}}, \qquad k = 1, 2.$$

We also have

$$S = \frac{v^*}{\kappa\theta_e} (0, 0, 1)^t (\theta_v/\theta_e, 0, 1),$$

$$(6.60) \qquad \alpha_k = \operatorname{tr}(ASA\mathcal{P}_k) = \left(\frac{1}{2}l_{01} - \frac{1}{2\beta_k}l_{02}\right) ASA(r_{01} - \beta_k r_{02})$$

$$= \frac{v^* p_e}{2\kappa\theta_e}(p - \theta_v/\theta_e) \equiv \alpha > 0, \qquad k = 1, 2.$$

Thus in (6.55),

$$H = \frac{1}{2} e^{-\alpha t} \sum_{k=1}^{2} \delta(x - \beta_k t) \begin{pmatrix} 1 & -1/\beta_k & 0 \\ -\beta_k & 1 & 0 \\ -\theta_v/\theta_e & \theta_v/(\theta_e\beta_k) & 0 \end{pmatrix}$$

with α and β_k, $k = 1, 2$, given by (6.60) and (6.59).

Next consider the linearization of the equations of magnetohydrodynamics (9.13) around a constant state $v = v^* > 0$, $u_1 = u_2 = u_3 = 0$, $B_2 = B_2^*$, $B_3 = B_3^*$, $e = e^* > 0$. We thus have system (6.50),

$$w_t + A w_x = B w_{xx},$$

where

$$w = (v - v^*, u_1, u_2, u_3, vB_2 - v^*B_2^*, vB_3 - v^*B_3^*, E - E^*)^t,$$

$$E^* = e^* + (B_2^{*2} + B_3^{*2})v^*/(2\mu_0),$$

$$A = \begin{pmatrix}
0 & -1 & 0 & 0 & 0 & 0 & 0 \\
c_1 & 0 & 0 & 0 & c_2B_2^* & c_2B_3^* & p_e \\
c_3B_2^* & 0 & 0 & 0 & -c_3 & 0 & 0 \\
c_3B_3^* & 0 & 0 & 0 & 0 & -c_3 & 0 \\
0 & 0 & -B_1^* & 0 & 0 & 0 & 0 \\
0 & 0 & 0 & -B_1^* & 0 & 0 & 0 \\
0 & P & -c_3B_2^*v^* & -c_3B_3^*v^* & 0 & 0 & 0
\end{pmatrix},$$

(6.61)
$$B = \begin{pmatrix}
0 & 0 & 0 & 0 & 0 & 0 & 0 \\
0 & \frac{\nu}{v^*} & 0 & 0 & 0 & 0 & 0 \\
0 & 0 & \frac{\mu}{v^*} & 0 & 0 & 0 & 0 \\
0 & 0 & 0 & \frac{\mu}{v^*} & 0 & 0 & 0 \\
-\frac{B_2^*}{\sigma\mu_0 v^{*2}} & 0 & 0 & 0 & \frac{1}{\sigma\mu_0 v^{*2}} & 0 & 0 \\
-\frac{B_3^*}{\sigma\mu_0 v^{*2}} & 0 & 0 & 0 & 0 & \frac{1}{\sigma\mu_0 v^{*2}} & 0 \\
a_1 & 0 & 0 & 0 & a_4B_2^* & a_4B_3^* & b
\end{pmatrix},$$

$$c_1 = p_v + \frac{1}{\mu_0}(B_2^{*2} + B_3^{*2})\left(\frac{p_e}{2} - \frac{1}{v^*}\right), \qquad c_2 = \frac{1}{\mu_0}\left(\frac{1}{v^*} - p_e\right),$$

$$c_3 = \frac{B_1^*}{\mu_0 v^*}, \qquad P = p + (B_2^{*2} + B_3^{*2})/(2\mu_0),$$

$$b = \frac{\kappa\theta_e}{v^*}, \qquad a_1 = \frac{\kappa}{v^*}\theta_v + (B_2^{*2} + B_3^{*2})\left(\frac{b}{2\mu_0} - \frac{1}{\sigma\mu_0^2 v^{*2}}\right),$$

$$a_4 = -\frac{b}{\mu_0} + \frac{1}{\sigma\mu_0^2 v^{*2}},$$

with p, θ_e, μ_0 and κ being positive constants, ν, μ and $1/\sigma$ nonnegative constants, and p_v, p_e, B_1^* and θ_v constants. Again we assume (6.52) for those constants related to thermodynamics. The matrix A has eigenvalues in nondecreasing order as

(6.62)
$$-c_f, -c_a, -c_s, 0, c_s, c_a, c_f,$$

where c_f, c_a, c_s are the fast, the Alfvén, and the slow wave speeds, respectively, given by

(6.63)
$$c_a^2 = \frac{B_1^{*2}}{\mu_0 v^*},$$
$$c_{f,s}^2 = \frac{1}{2}\left[d^2 \pm \sqrt{d^4 + 4\bar{p}_v B_1^{*2}/(\mu_0 v^*)}\right],$$
$$d^2 = -\bar{p}_v + (B_1^{*2} + B_2^{*2} + B_3^{*2})/(\mu_0 v^*),$$

with $\bar{p}_v$ defined in (6.52). In the following cases some of the eigenvalues coincide:

Case A. When $B_1^* = 0$, $c_s = c_a = 0$, thus $\lambda = 0$ is an eigenvalue of multiplicity 5.

Case B. When $B_1^* \neq 0$ and $B_2^* = B_3^* = 0$, we have $c_f^2 = \max(-\bar{p}_v, c_a^2)$, and $c_s^2 = \min(-\bar{p}_v, c_a^2)$. If $-\bar{p}_v \neq c_a^2$, either $c_f^2 = c_a^2$ or $c_s^2 = c_a^2$, hence $\lambda = \pm c_a$ are eigenvalues of multiplicity 2. If $-\bar{p}_v = c_a^2$, the multiplicity of $\lambda = \pm c_a$ is 3.

Set

$$r(c) = \big(1, -c, cB_1^* B_2^*/\rho, cB_1^* B_3^*/\rho, -B_1^{*2} B_2^*/\rho, -B_1^{*2} B_3^*/\rho,$$
$$- p - (B_2^{*2} + B_3^{*2})(\mu_0 v^* c^2 + B_1^{*2})/(2\mu_0 \rho)\big)^t,$$

$$(6.64) \quad l(c) = \big(2 + 2B_1^{*2}(B_2^{*2} + B_3^{*2})/\rho^2\big)^{-1} \big(1 - Pp_e/c^2, -1/c, B_1^* B_2^*/(c\rho), B_1^* B_3^*/(c\rho),$$
$$p_e B_2^*/(\mu_0 c^2) - B_2^*/\rho, p_e B_3^*/(\mu_0 c^2) - B_3^*/\rho, -p_e/c^2\big),$$

$$\rho = \rho(c) = \mu_0 v^* c^2 - B_1^{*2}.$$

We now write down the eigenvectors of A satisfying (6.2) for each case.

Case A. $B_1^* = 0$. The matrix A has eigenvalues

$$(6.65) \qquad \lambda_1 = -c_f, \qquad \lambda_2 = 0, \qquad \lambda_3 = c_f,$$

with $m_1 = m_3 = 1$, $m_2 = 5$. The right eigenvectors are

$$r_{11} = r(-c_f), \qquad r_{31} = r(c_f),$$
$$r_{21} = (1, 0, 0, 0, 0, 0, -c_1/p_e)^t,$$
$$(6.66a) \qquad r_{22} = (0, 0, 1, 0, 0, 0, 0)^t,$$
$$r_{23} = (0, 0, 0, 1, 0, 0, 0)^t,$$
$$r_{24} = (0, 0, 0, 0, 1, 0, -c_2 B_2^*/p_e)^t,$$
$$r_{25} = (0, 0, 0, 0, 0, 1, -c_2 B_3^*/p_e)^t.$$

The left eigenvectors are

$$l_{11} = l(-c_f), \qquad l_{31} = l(c_f),$$
$$l_{21} = \left(-\bar{p}_v + \frac{B_2^{*2} + B_3^{*2}}{\mu_0 v^*}\right)^{-1} (Pp_e, 0, 0, 0, c_2 B_2^*, c_2 B_3^*, p_e),$$

(6.66*b*) $l_{22} = (0, 0, 1, 0, 0, 0, 0),$

$l_{23} = (0, 0, 0, 1, 0, 0, 0),$

$l_{24} = (0, 0, 0, 0, 1, 0, 0),$

$l_{25} = (0, 0, 0, 0, 0, 1, 0).$

Case B1. $B_1^* \neq 0$, $B_2^* = B_3^* = 0$, and $-\bar{p}_v \neq c_a^2$. The matrix A has eigenvalues

(6.67) $\lambda_1 = -\sqrt{-\bar{p}_v}, \quad \lambda_2 = -c_a, \quad \lambda_3 = 0, \quad \lambda_4 = c_a, \quad \lambda_5 = \sqrt{-\bar{p}_v},$

with $m_1 = m_3 = m_5 = 1$, $m_2 = m_4 = 2$. Here λ_1 and λ_5 are the fast waves or the slow waves. The right eigenvectors are

$$r_{11} = (1, \sqrt{-\bar{p}_v}, 0, 0, 0, 0, -p)^t,$$

$$r_{51} = (1, -\sqrt{-\bar{p}_v}, 0, 0, 0, 0, -p)^t,$$

$$r_{31} = (1, 0, 0, 0, 0, 0, -p_v/p_e)^t,$$

(6.68*a*) $$r_{21} = (0, 0, \operatorname{sign}(B_1^*)/\sqrt{\mu_0 v^*}, 0, 1, 0, 0)^t,$$

$$r_{22} = (0, 0, 0, \operatorname{sign}(B_1^*)/\sqrt{\mu_0 v^*}, 0, 1, 0)^t,$$

$$r_{41} = (0, 0, -\operatorname{sign}(B_1^*)/\sqrt{\mu_0 v^*}, 0, 1, 0, 0)^t,$$

$$r_{42} = (0, 0, 0, -\operatorname{sign}(B_1^*)/\sqrt{\mu_0 v^*}, 0, 1, 0)^t.$$

The left eigenvectors are

$$l_{11} = \frac{1}{2}(p_v/\bar{p}_v, 1/\sqrt{-\bar{p}_v}, 0, 0, 0, 0, p_e/\bar{p}_v),$$

$$l_{51} = \frac{1}{2}(p_v/\bar{p}_v, -1/\sqrt{-\bar{p}_v}, 0, 0, 0, 0, p_e/\bar{p}_v),$$

$$l_{31} = (-pp_e/\bar{p}_v, 0, 0, 0, 0, 0, -p_e/\bar{p}_v),$$

(6.68*b*) $$l_{21} = \frac{1}{2}(0, 0, \sqrt{\mu_0 v^*} \operatorname{sign}(B_1^*), 0, 1, 0, 0),$$

$$l_{22} = \frac{1}{2}(0, 0, 0, \sqrt{\mu_0 v^*} \operatorname{sign}(B_1^*), 0, 1, 0),$$

$$l_{41} = \frac{1}{2}(0, 0, -\sqrt{\mu_0 v^*} \operatorname{sign}(B_1^*), 0, 1, 0, 0),$$

$$l_{42} = \frac{1}{2}(0, 0, 0, -\sqrt{\mu_0 v^*} \operatorname{sign}(B_1^*), 0, 1, 0).$$

Case B2. $B_1^* \neq 0$, $B_2^* = B_3^* = 0$, and $-\bar{p}_v = c_a^2$. The matrix A has eigenvalues

$$(6.69) \qquad\qquad \lambda_1 = -c_a, \qquad \lambda_2 = 0, \qquad \lambda_3 = c_a,$$

with $m_1 = m_3 = 3$, $m_2 = 1$. The eigenvectors are the same as in Case B1.

Case C. $B_1^* \neq 0$, and $B_2^{*2} + B_3^{*2} \neq 0$. The matrix A has eigenvalues λ_i given by (6.62), with $m_i = 1$, $i = 1, \ldots, 7$. The right Eigenvectors are

$$r_1 = r(-c_f), \qquad r_7 = r(c_f), \qquad r_3 = r(-c_s), \qquad r_5 = r(c_s),$$

$$r_4 = \left(1, 0, 0, 0, B_2^*, B_3^*, -\frac{p_v}{p_e} + \frac{B_2^{*2} + B_3^{*2}}{2\mu_0}\right)^t,$$

$$(6.70a)$$

$$r_2 = \left(0, 0, \frac{B_3^* \operatorname{sign}(B_1^*)}{\sqrt{\mu_0 v^*}}, -\frac{B_2^* \operatorname{sign}(B_1^*)}{\sqrt{\mu_0 v^*}}, B_3^*, -B_2^*, 0\right)^t,$$

$$r_6 = \left(0, 0, -\frac{B_3^* \operatorname{sign}(B_1^*)}{\sqrt{\mu_0 v^*}}, \frac{B_2^* \operatorname{sign}(B_1^*)}{\sqrt{\mu_0 v^*}}, B_3^*, -B_2^*, 0\right)^t.$$

The left eigenvectors are

$$l_1 = l(-c_f), \qquad l_7 = l(c_f), \qquad l_3 = l(-c_s), \qquad l_5 = l(c_s),$$

$$l_4 = -\frac{p_e}{\bar{p}_v}\left(P, 0, 0, 0, -\frac{B_2^*}{\mu_0}, -\frac{B_3^*}{\mu_0}, 1\right),$$

$$(6.70b)$$

$$l_2 = \frac{1}{2(B_2^{*2} + B_3^{*2})}\left(0, 0, \sqrt{\mu_0 v^*} B_3^* \operatorname{sign}(B_1^*), -\sqrt{\mu_0 v^*} B_2^* \operatorname{sign}(B_1^*), B_3^*, -B_2^*, 0\right),$$

$$l_6 = \frac{1}{2(B_2^{*2} + B_3^{*2})}\left(0, 0, -\sqrt{\mu_0 v^*} B_3^* \operatorname{sign}(B_1^*), \sqrt{\mu_0 v^*} B_2^* \operatorname{sign}(B_1^*), B_3^*, -B_2^*, 0\right).$$

Assumption 6.1 is implied by Assumption 2.1 for the nonlinear system, which will be verified in Section 9. Notice that we have assumed $\kappa > 0$. By the above expressions for the eigenvectors of A, we see that Assumption 6.2 is satisfied for the following choices of parameters:

   (i) $\nu > 0$, $\mu > 0$, and $1/\sigma > 0$;
   (ii) $\nu = \mu = 0$, and $1/\sigma > 0$, while $B_1^* \neq 0$;
   (iii) $\nu > 0$, $\mu > 0$, and $1/\sigma = 0$, while $B_1^* \neq 0$.

Now we apply Theorem 6.2 to (6.50), (6.61). The Green's function of the linearized equations of magnetohydrodynamics is

$$(6.71) \qquad G(x,t) = G^*(x,t) + O(1)(t+1)^{-\frac{1}{2}} t^{-\frac{1}{2}} \sum_j e^{-\frac{(x-\lambda_j t)^2}{Ct}} + H,$$

where G^* is the Green's function of a diagonalizable parabolic system

$$(6.72) \qquad\qquad w_t + Aw_x = B^* w_{xx},$$

the λ_j are the eigenvalues of A given by (6.62), (6.63), and $H = \sum_{k=1}^{m'} e^{-\alpha_k t} \delta(x - \beta_k t) \mathcal{P}_k$. The matrix B^* is given by (6.5). Notice that we have computed all the eigenvectors of A. Hence the computation of G^* is straightforward. For example, in Case C we have

$$(6.73) \qquad\qquad G^*(x,t) = \sum_{j=1}^{7} \frac{1}{\sqrt{4\pi\mu_j t}} e^{-\frac{(x-\lambda_j t)^2}{4\mu_j t}} r_j l_j,$$

where

$$\mu_1 = \mu_7 = \mu(c_f), \qquad \mu_3 = \mu_5 = \mu(c_s),$$

$$\mu(c) = l(c) Br(c) = \left[2 + 2B_1^{*2}(B_2^{*2} + B_3^{*2})/\rho^2\right]^{-1}$$

$$(6.74) \qquad\qquad \cdot \left[\frac{\kappa}{v^* c^2} p_e(\theta_e p - \theta_v) + \frac{\nu}{v^*} + \frac{\mu}{v^* \rho^2} B_1^{*2}(B_2^{*2} + B_3^{*2}) + \frac{c^2}{\sigma v^* \rho^2}(B_2^{*2} + B_3^{*2})\right],$$

$$\mu_4 = l_4 Br_4 = \frac{\kappa \theta_e \tilde{p}_v}{v^* \bar{p}_v},$$

$$\mu_2 = \mu_6 = l_2 Br_2 = l_6 Br_6 = \frac{\mu}{2v^*} + \frac{1}{2\sigma\mu_0 v^{*2}}.$$

The other cases are somewhat tedious. We now concentrate on obtaining H. Denote the eigenvalues of B as ν_i:

$$(6.75) \qquad \nu_1 = 0, \quad \nu_2 = \frac{\nu}{v^*}, \quad \nu_3 = \nu_4 = \frac{\mu}{v^*}, \quad \nu_5 = \nu_6 = \frac{1}{\sigma\mu_0 v^{*2}}, \quad \nu_7 = \frac{\kappa\theta_e}{v^*}.$$

The corresponding right eigenvectors of B are

$$\mathfrak{p}_1 = \left(1, 0, 0, 0, B_2^*, B_3^*, (B_2^{*2} + B_3^{*2})/(2\mu_0) - \theta_v/\theta_e\right)^t,$$

$$\mathfrak{p}_2 = (0, 1, 0, 0, 0, 0, 0)^t,$$

$$\mathfrak{p}_3 = (0, 0, 1, 0, 0, 0, 0)^t,$$

$$(6.76a) \qquad \mathfrak{p}_4 = (0, 0, 0, 1, 0, 0, 0)^t,$$

$$\mathfrak{p}_5 = (0, 0, 0, 0, 1, 0, B_2^*/\mu_0)^t,$$

$$\mathfrak{p}_6 = (0, 0, 0, 0, 0, 1, B_3^*/\mu_0)^t,$$

$$\mathfrak{p}_7 = (0, 0, 0, 0, 0, 0, 1)^t,$$

and the left eigenvectors are

$$\mathfrak{q}_1 = (1,0,0,0,0,0,0),$$
$$\mathfrak{q}_2 = (0,1,0,0,0,0,0),$$
$$\mathfrak{q}_3 = (0,0,1,0,0,0,0),$$

(6.76b) $$\mathfrak{q}_4 = (0,0,0,1,0,0,0),$$
$$\mathfrak{q}_5 = (-B_2^*,0,0,0,1,0,0),$$
$$\mathfrak{q}_6 = (-B_3^*,0,0,0,0,1,0),$$
$$\mathfrak{q}_7 = \big((B_2^{*2}+B_3^{*2})/(2\mu_0)+\theta_v/\theta_e,0,0,0,-B_2^*/\mu_0,-B_3^*/\mu_0,1\big),$$

satisfying

$$\mathfrak{q}_i\mathfrak{p}_j = \delta_{ij}, \qquad i,j = 1,\ldots,7.$$

Case (i). $\nu > 0$, $\mu > 0$, and $1/\sigma > 0$. In this case $m' = m = 1$ by Theorem 6.2, and

$$\mathcal{P} = \mathfrak{p}_1\mathfrak{q}_1, \qquad \beta = \mathfrak{q}_1 A\mathfrak{p}_1 = 0.$$

By Remark 6.4,

$$S = \sum_{j=2}^{7} \frac{1}{\nu_j}\mathfrak{p}_j\mathfrak{q}_j,$$
$$\alpha = \mathrm{tr}(ASA\mathcal{P}) = \mathfrak{q}_1 ASA\mathfrak{p}_1 = -v^*\tilde{p}_v/\nu,$$

where $\tilde{p}_v$ is defined in (6.52). Hence

(6.77) $$H = e^{v^*\tilde{p}_v t/\nu}\delta(x)\mathfrak{p}_1\mathfrak{q}_1.$$

Case (ii). $\nu = \mu = 0$, $1/\sigma > 0$, and $B_1^* \neq 0$. In this case $m = 4$, and $Q_0 = \sum_{j=1}^{4} \mathfrak{p}_j\mathfrak{q}_j$. Direct computation shows that

(6.78) $$Q_0 A Q_0 = (\mathfrak{p}_1,\mathfrak{p}_2,\mathfrak{p}_3,\mathfrak{p}_4)\begin{pmatrix} 0 & -1 & 0 & 0 \\ \tilde{p}_v & 0 & 0 & 0 \\ 0 & 0 & 0 & 0 \\ 0 & 0 & 0 & 0 \end{pmatrix}\begin{pmatrix} \mathfrak{q}_1 \\ \mathfrak{q}_2 \\ \mathfrak{q}_3 \\ \mathfrak{q}_4 \end{pmatrix} = \beta_1\tilde{p}_1\tilde{q}_1 + \beta_2\tilde{p}_2\tilde{q}_2,$$

where

$$\beta_{1,2} = \mp\sqrt{-\tilde{p}_v},$$

(6.79)

$$\tilde{\mathfrak{p}}_{1,2} = \mathfrak{p}_1 - \beta_{1,2}\mathfrak{p}_2, \qquad \tilde{\mathfrak{q}}_{1,2} = \frac{1}{2}\mathfrak{q}_1 - \frac{1}{2\beta_{1,2}}\mathfrak{q}_2.$$

It is easy to see that the right-hand side of (6.78) is the spectral representation. Hence with the notations in Remark 6.4,

(6.80)

$$\mathcal{P}_k = \mathcal{Q}_k = \tilde{\mathfrak{p}}_k\tilde{\mathfrak{q}}_k, \qquad k = 1, 2,$$

$$\beta_3 = 0, \qquad \mathcal{Q}_3 = \mathcal{Q}_0 - \mathcal{Q}_1 - \mathcal{Q}_2 = \mathfrak{p}_3\mathfrak{q}_3 + \mathfrak{p}_4\mathfrak{q}_4.$$

Notice that

$$S = \sum_{j=5}^{7} \frac{1}{\nu_j}\mathfrak{p}_j\mathfrak{q}_j.$$

We then have for $k = 1, 2$,

(6.81)
$$\alpha_k = \tilde{\mathfrak{q}}_k A S A \tilde{\mathfrak{p}}_k = \frac{\sigma v^*}{2}(B_2^{*2} + B_3^{*2}) + \frac{v^* p_e(\theta_e p - \theta_v)}{2\kappa\theta_e^2} \equiv \alpha > 0$$

by (6.52). By (6.80) and direct computation,

$$\mathcal{Q}_3 A S A \mathcal{Q}_3 = \frac{c_3 B_1^*}{\nu_5}(\mathfrak{p}_3\mathfrak{q}_3 + \mathfrak{p}_4\mathfrak{q}_4) = \sigma v^* B_1^{*2}\mathcal{Q}_3.$$

Therefore,

$$\alpha_3 = \sigma v^* B_1^{*2}, \qquad \mathcal{P}_3 = \mathcal{Q}_3.$$

In conclusion,

(6.82) $H = e^{-\alpha t}\left[\delta(x + t\sqrt{-\tilde{p}_v})\tilde{\mathfrak{p}}_1\tilde{\mathfrak{q}}_1 + \delta(x - t\sqrt{-\tilde{p}_v})\tilde{\mathfrak{p}}_2\tilde{\mathfrak{q}}_2\right] + e^{-\sigma v^* B_1^{*2}t}\delta(x)(\mathfrak{p}_3\mathfrak{q}_3 + \mathfrak{p}_4\mathfrak{q}_4),$

where α and $\tilde{\mathfrak{p}}_k$, $\tilde{\mathfrak{q}}_k$, $k = 1, 2$, are defined in (6.81) and (6.79).

Case (iii). $\nu > 0$, $\mu > 0$, $1/\sigma = 0$ and $B_1^* \neq 0$. In this case $m = 3$, and

$$\mathcal{Q}_0 = \mathfrak{p}_1\mathfrak{q}_1 + \mathfrak{p}_5\mathfrak{q}_5 + \mathfrak{p}_6\mathfrak{q}_6.$$

Direct computation shows that $\mathcal{Q}_0 A \mathcal{Q}_0 = 0$. Hence $\beta = 0$ and $\mathcal{Q} = \mathcal{Q}_0$. Notice that

$$S = \frac{1}{\nu_2}\mathfrak{p}_2\mathfrak{q}_2 + \frac{1}{\nu_3}(\mathfrak{p}_3\mathfrak{q}_3 + \mathfrak{p}_4\mathfrak{q}_4) + \frac{1}{\nu_7}\mathfrak{p}_7\mathfrak{q}_7.$$

We now compute α_k. We have

(6.83)

$$QASAQ = (\mathfrak{p}_1, \mathfrak{p}_5, \mathfrak{p}_6) \left\{ \frac{1}{\nu_2} \begin{pmatrix} -1 \\ B_2^* \\ B_3^* \end{pmatrix} \left(\tilde{p}_v, \frac{B_2^*}{\mu_0 v^*}, \frac{B_3^*}{\mu_0 v^*} \right) + \frac{1}{\nu_3} \begin{pmatrix} 0 & & \\ & \frac{B_1^{*2}}{\mu_0 v^*} & \\ & & \frac{B_1^{*2}}{\mu_0 v^*} \end{pmatrix} \right\} \begin{pmatrix} \mathfrak{q}_1 \\ \mathfrak{q}_5 \\ \mathfrak{q}_6 \end{pmatrix}.$$

If $B_2^{*2} + B_3^{*2} = 0$, then

$$QASAQ = -\frac{\tilde{p}_v v^*}{\nu} \mathfrak{p}_1 \mathfrak{q}_1 + \frac{B_1^{*2}}{\mu \mu_0} (\mathfrak{p}_5 \mathfrak{q}_5 + \mathfrak{p}_6 \mathfrak{q}_6).$$

The right-hand side is the spectral representation $\alpha_1 \mathcal{P}_1 + \alpha_2 \mathcal{P}_2$. Therefore,

(6.84)
$$H = \delta(x) \left[e^{\frac{\tilde{p}_v v^* t}{\nu}} \mathfrak{p}_1 \mathfrak{q}_1 + e^{-\frac{B_1^{*2} t}{\mu \mu_0}} (\mathfrak{p}_5 \mathfrak{q}_5 + \mathfrak{p}_6 \mathfrak{q}_6) \right].$$

If $B_2^{*2} + B_3^{*2} \neq 0$, (6.83) has the spectral representation

$$QASAQ = \alpha_1 \mathcal{P}_1 + \alpha_2 \mathcal{P}_2 + \alpha_3 \mathcal{P}_3,$$

where

(6.85)
$$\alpha_1 = \frac{B_1^{*2}}{\mu_0 \mu}, \qquad \alpha_{2,3} = \frac{1}{2} \left[\omega^2 \pm \sqrt{\omega^4 + 4\tilde{p}_v B_1^{*2} v^* / (\mu_0 \nu \mu)} \right] \neq \alpha_1,$$

$$\omega^2 = -\frac{\tilde{p}_v v^*}{\nu} + \frac{B_2^{*2} + B_3^{*2}}{\nu \mu_0} + \frac{B_1^{*2}}{\mu \mu_0},$$

$$\mathcal{P}_1 = [B_3^{*2} \mathfrak{p}_5 \mathfrak{q}_5 + B_2^{*2} \mathfrak{p}_6 \mathfrak{q}_6 - B_2^* B_3^* (\mathfrak{p}_6 \mathfrak{q}_5 + \mathfrak{p}_5 \mathfrak{q}_6)] / (B_2^{*2} + B_3^{*2}),$$

$$\mathcal{P}_{2,3} = \left[(\alpha_{2,3} - \alpha_1)^2 - \frac{\alpha_{2,3}^2}{\mu_0 v^* \tilde{p}_v} (B_2^{*2} + B_3^{*2}) \right]^{-1} [(\alpha_{2,3} - \alpha_1) \mathfrak{p}_1 - \alpha_{2,3} B_2^* \mathfrak{p}_5 - \alpha_{2,3} B_3^* \mathfrak{p}_6]$$
$$\left[(\alpha_{2,3} - \alpha_1) \mathfrak{q}_1 + \frac{\alpha_{2,3} B_2^*}{\mu_0 v^* \tilde{p}_v} \mathfrak{q}_5 + \frac{\alpha_{2,3} B_3^*}{\mu_0 v^* \tilde{p}_v} \mathfrak{q}_6 \right].$$

Therefore,

(6.86)
$$H = \delta(x) \left[e^{-\alpha_1 t} \mathcal{P}_1 + e^{-\alpha_2 t} \mathcal{P}_2 + e^{-\alpha_3 t} \mathcal{P}_3 \right],$$

where α_i and $\mathcal{P}_i$, $i = 1, 2, 3$, are given by (6.85).

7. ENERGY ESTIMATE

In this section we present the energy estimate, which is due to Kawashima, for the Cauchy Problem (2.1), (2.2) of viscous conservation laws. The reader is referred to [Ka] and the references therein.

Theorem 7.1. *Suppose that Assumptions 2.1–2.3 hold. Let $m \geq 2$ be an integer and assume that $u_0 \in H^m$. If $\|u_0\|_{H^m}$ is small, then the initial value problem (2.1), (2.2) has a unique global solution $u(x,t)$ satisfying*

(7.1)
$$u^{(1)} \in C^0([0,\infty); H^m) \cap C^1([0,\infty); H^{m-1}),$$
$$u^{(2)} \in C^0([0,\infty); H^m) \cap C^1([0,\infty); H^{m-2}),$$
$$u_x^{(1)} \in L^2([0,\infty); H^{m-1}), \qquad u_x^{(2)} \in L^2([0,\infty); H^m),$$

where $u^{(1)}(x,t)$ and $u^{(2)}(x,t)$ are the orthogonal projections of $\tilde{u}(x,t) = f_0^{-1}(u(x,t))$ on $\mathcal{N}$ and $\mathcal{N}^{\perp}$ respectively; f_0 is the mapping in Assumption 2.2; and $\mathcal{N}$ is the null space of $B(f_0(\tilde{u}))f_0'(\tilde{u})$. The solution satisfies

(7.2)
$$\sup_{0 \leq \tau \leq t} \|u(\cdot, \tau)\|_{H^m}^2 + \int_0^t \left(\|u_x(\cdot, \tau)\|_{H^{m-1}}^2 + \|u_x^{(2)}(\cdot, \tau)\|_{H^m}^2 \right) d\tau \leq C\|u_0\|_{H^m}^2$$

for all $t \geq 0$, where C is a constant.

Before proving Theorem 7.1, we restate (2.1), (2.2) and Assumptions 2.1–2.3 here for convenience. We consider the Cauchy problem

(7.3)
$$u_t + f(u)_x = (B(u)u_x)_x,$$

(7.4)
$$u(x,0) = u_0(x),$$

Here (7.3) satisfies the following conditions:

(i) There exist an entropy pair $U(u)$ and $F(u)$. Set

(7.5)
$$A_0(u) = \nabla^2 U(u), \qquad A(u) = f'(u).$$

Then $(\nabla U)A = \nabla F$, A_0 is positive definite, $A_0 A$ is symmetric, and $A_0 B$ is symmetric semi-positive definite.

(ii) For small u, $B(u) \neq 0$. There exists a smooth one-to-one mapping

$$(7.6) \qquad\qquad u = f_0(\tilde{u}), \qquad f_0(0) = 0,$$

such that the null space $\mathcal{N}$ of

$$(7.7) \qquad\qquad \tilde{B}(\tilde{u}) \equiv B(f_0(\tilde{u}))f_0'(\tilde{u})$$

is independent of $\tilde{u}$. Moreover, $\mathcal{N}^\perp$ is invariant under $f_0'(\tilde{u})^t A_0(f_0(\tilde{u}))$, and $\tilde{B}(\tilde{u})$ maps $\mathbb{R}^n$ into $\mathcal{N}^\perp$.

(iii) Any eigenvector of $A(0)$ is not in the null space of $B(0)$. By Lemma 6.6, this implies that there exists a constant matrix $K \in \mathbb{R}^{n \times n}$, such that $KA_0(0)$ is skew symmetric and $\{\frac{1}{2}[KA_0A + (KA_0A)^t] + A_0B\}(0)$ is positive definite.

Assumption (ii) allows us to consider such a system (7.3) whose viscosity matrix $B(u)$ has a null space depending on u, but after transform (7.6) the system for $\tilde{u}$,

$$f_0(\tilde{u})_t + f(f_0(\tilde{u}))_x = (\tilde{B}(\tilde{u})\tilde{u}_x)_x,$$

has a new viscosity matrix $\tilde{B}(\tilde{u})$ whose null space is independent of $\tilde{u}$. See examples in Section 9.

Since the local existence is standard [K], to prove Theorem 7.1 it is sufficient to obtain the following a priori estimate:

Lemma 7.2. *Assume conditions (i)–(iii). Let $m \geq 2$ be an integer and let $T > 0$ be a constant. Let $u(x,t)$ be a solution to (7.3), (7.4) satisfying (7.1) with $[0, \infty)$ replaced by $[0, T]$. Set*

$$(7.8) \qquad N_m(t)^2 = \sup_{0 \leq \tau \leq t} \|u(\cdot, \tau)\|_{H^m}^2 + \int_0^t \left(\|u_x(\cdot, \tau)\|_{H^{m-1}}^2 + \|u_x^{(2)}(\cdot, \tau)\|_{H^m}^2 \right) d\tau.$$

If $N_m(T)$ is bounded by a small positive constant independent of T, then $N_m(T) \leq C\|u_0\|_{H^m}$, where C is a constant independent of T.

Proof. To simplify the notation we denote $\| \cdot \|_{L^2}$ by $\| \cdot \|$. Without loss of generality we assume that $U(0) = F(0) = 0$ and $f(0) = 0$. Set

$$\mathcal{E}(u) = U(u) - \nabla U(0)u.$$

Clearly $\mathcal{E}$ is equivalent to $|u|^2$ for small u since $A_0 = \nabla^2 U$ is positive definite. By condition (i) and (7.3),

$$\mathcal{E}(u)_t + (F(u) - \nabla U(0)f)_x = (\nabla U(u) - \nabla U(0))(B(u)u_x)_x.$$

Integrate this equality over $\mathbb{R} \times [0,t]$ for $0 \le t \le T$. We have

$$(7.9) \qquad c_1\|u\|^2(t) + \int_0^t \int_{-\infty}^\infty (\nabla U(u))_x B(u)u_x \, dxd\tau \le c_2\|u_0\|^2,$$

where c_1 and c_2 are positive constants. By conditions (i) and (ii) the integral in (7.9) is

$$\int_0^t \int_{-\infty}^\infty u_x^t A_0(u)B(u)u_x \, dxd\tau$$

$$= \int_0^t \int_{-\infty}^\infty (u_x^{(2)})^t f_0'(\tilde{u})^t A_0(u)B(u)f_0'(\tilde{u})u_x^{(2)} \, dxd\tau \ge c_3 \int_0^t \|u_x^{(2)}\|^2(\tau)\, d\tau,$$

where $c_3 > 0$ is a constant and $u^{(2)}$ is the projection of $\tilde{u}$ on $\mathcal{N}^\perp$. Thus (7.9) implies that

$$(7.10) \qquad \|u\|^2(t) + \int_0^t \|u_x^{(2)}\|^2(\tau)\, d\tau \le C\|u_0\|^2$$

for $t \in [0,T]$, where C is a constant independent of T. Here and later, $N_m(T)$ is assumed to be bounded by a small positive constant independent of T.

We then estimate the L^2 norm of derivatives. Applying $\partial^j/\partial x^j$, $1 \le j \le m$, to (7.3), multiplying it by $A_0(u)$ from the left, and taking the inner product $\langle \cdot, \cdot \rangle$ with $\partial^j u/\partial x^j$, we have

$$\frac{1}{2}\left\langle \frac{\partial^j u}{\partial x^j}, A_0(u)\frac{\partial^j u}{\partial x^j}\right\rangle_t + \left\langle \frac{\partial^j u}{\partial x^j}, A_0(u)\frac{\partial^j}{\partial x^j}(A(u)u_x)\right\rangle$$

$$= \left\langle \frac{\partial^j u}{\partial x^j}, A_0(u)\frac{\partial^j}{\partial x^j}(B(u)u_x)_x\right\rangle + \frac{1}{2}\left\langle \frac{\partial^j u}{\partial x^j}, A_0(u)_t\frac{\partial^j u}{\partial x^j}\right\rangle.$$

Integrate this equation over $\mathbb{R} \times [0,t]$ for $0 \le t \le T$. By integration by parts and condition (i) we have

$$(7.11)$$
$$c_1\left\|\frac{\partial^j u}{\partial x^j}\right\|^2(t) + \int_0^t \int_{-\infty}^\infty \left\langle \frac{\partial^{j+1}u}{\partial x^{j+1}}, A_0(u)\frac{\partial^j}{\partial x^j}(B(u)u_x)\right\rangle dxd\tau$$

$$\le c_2\|u_0^{(j)}\|^2 - \int_0^t \int_{-\infty}^\infty \left\langle \frac{\partial^j u}{\partial x^j}, A_0(u)\left[\frac{\partial^j}{\partial x^j}(A(u)u_x) - A(u)\frac{\partial^{j+1}u}{\partial x^{j+1}}\right]\right\rangle dxd\tau$$

$$+ \frac{1}{2}\int_0^t \int_{-\infty}^\infty \left\langle \frac{\partial^j u}{\partial x^j}, (A_0(u)A(u))_x\frac{\partial^j u}{\partial x^j}\right\rangle dxd\tau$$

$$- \int_0^t \int_{-\infty}^\infty \left\langle \frac{\partial^j u}{\partial x^j}, A_0(u)_x\frac{\partial^j}{\partial x^j}(B(u)u_x)\right\rangle dxd\tau + \frac{1}{2}\int_0^t \int_{-\infty}^\infty \left\langle \frac{\partial^j u}{\partial x^j}, A_0(u)_t\frac{\partial^j u}{\partial x^j}\right\rangle dxd\tau,$$

where c_1 and c_2 are again positive constants. Let P be the projection operator on $\mathcal{N}^\perp$, which is a constant matrix. By conditions (i) and (ii) the second term on the left-hand side of (7.11) is

$$\int_0^t \int_{-\infty}^\infty \left\langle \frac{\partial^j}{\partial x^j}(f_0'(\tilde{u})\tilde{u}_x), A_0(u)\frac{\partial^j}{\partial x^j}(P\tilde{B}(\tilde{u})u_x^{(2)}) \right\rangle dx d\tau$$

$$= \int_0^t \int_{-\infty}^\infty \left\langle \frac{\partial^j \tilde{u}_x}{\partial x^j}, f_0'(\tilde{u})^t A_0(u)P\frac{\partial^j}{\partial x^j}(\tilde{B}(\tilde{u})u_x^{(2)}) \right\rangle dx d\tau + O(1)N_m(t)^3$$

$$= \int_0^t \int_{-\infty}^\infty \left\langle \frac{\partial^j u_x^{(2)}}{\partial x^j}, f_0'(\tilde{u})^t A_0(u)\tilde{B}(\tilde{u})\frac{\partial^j u_x^{(2)}}{\partial x^j} \right\rangle dx d\tau + O(1)N_m(t)^3$$

$$\geq c_3 \int_0^t \left\| \frac{\partial^j u_x^{(2)}}{\partial x^j} \right\|^2 (\tau)\, d\tau - CN_m(t)^3,$$

where $c_3 > 0$ is a constant, and we have noticed $m \geq 2$, $1 \leq j \leq m$. Thus (7.11) is reduced to

$$(7.12) \qquad \left\| \frac{\partial^j u}{\partial x^j} \right\|^2 (t) + \int_0^t \left\| \frac{\partial^j u_x^{(2)}}{\partial x^j} \right\|^2 (\tau)\, d\tau \leq C\|u_0\|_{H^m}^2 + CN_m(t)^3, \qquad 1 \leq j \leq m,$$

for $t \in [0, T]$.

It is sufficient to obtain an estimate for the $L^2([0,T]; H^{m-1})$ norm of u_x. Rewrite (7.3) as

$$(7.13) \qquad\qquad\qquad u_t + A(0)u_x = h,$$

where

$$(7.14) \qquad\qquad\qquad h = (A(0) - A(u))u_x + (\tilde{B}(\tilde{u})u_x^{(2)})_x.$$

From (7.13) we have for $0 \leq j \leq m-1$, $0 \leq t \leq T$,

$$(7.15)$$
$$\int_0^t \int_{-\infty}^\infty \left\langle \frac{\partial^{j+1}u}{\partial x^{j+1}}, KA_0(0)\frac{\partial^{j+1}u}{\partial x^j \partial t} \right\rangle dx d\tau + \int_0^t \int_{-\infty}^\infty \left\langle \frac{\partial^{j+1}u}{\partial x^{j+1}}, KA_0(0)A(0)\frac{\partial^{j+1}u}{\partial x^{j+1}} \right\rangle dx d\tau$$
$$= \int_0^t \int_{-\infty}^\infty \left\langle \frac{\partial^{j+1}u}{\partial x^{j+1}}, KA_0(0)\frac{\partial^j h}{\partial x^j} \right\rangle dx d\tau,$$

where K is as in condition (iii). For the first term on the left, we integrate by parts with respect to x and then with respect to t, which gives

$$\int_0^t \int_{-\infty}^\infty \left\langle \frac{\partial^{j+1}u}{\partial x^{j+1}}, KA_0(0)\frac{\partial^{j+1}u}{\partial x^j \partial t} \right\rangle dx d\tau = -\int_{-\infty}^\infty \left\langle \frac{\partial^j u}{\partial x^j}, KA_0(0)\frac{\partial^{j+1}u}{\partial x^{j+1}} \right\rangle (x,t)\, dx$$
$$+ \int_{-\infty}^\infty \left\langle u_0^{(j)}, KA_0(0)u_0^{(j+1)} \right\rangle dx + \int_0^t \int_{-\infty}^\infty \left\langle \frac{\partial^{j+1}u}{\partial x^j \partial t}, KA_0(0)\frac{\partial^{j+1}u}{\partial x^{j+1}} \right\rangle dx d\tau.$$

Notice that $KA_0(0)$ is skew symmetric. Therefore,

$$
\begin{aligned}
(7.16) \quad & \int_0^t \int_{-\infty}^{\infty} \left\langle \frac{\partial^{j+1} u}{\partial x^{j+1}}, KA_0(0) \frac{\partial^{j+1} u}{\partial x^j \partial t} \right\rangle dx d\tau \\
& = -\frac{1}{2} \int_{-\infty}^{\infty} \left\langle \frac{\partial^j u}{\partial x^j}, KA_0(0) \frac{\partial^{j+1} u}{\partial x^{j+1}} \right\rangle (x,t) \, dx + \frac{1}{2} \int_{-\infty}^{\infty} \left\langle u_0^{(j)}, KA_0(0) u_0^{(j+1)} \right\rangle dx \\
& \geq -C \|u\|_{H^m}^2(t) - C \|u_0\|_{H^m}^2
\end{aligned}
$$

since $0 \leq j \leq m-1$. For the second term on the left-hand side of (7.15), we decompose $KA_0(0)A(0)$ into symmetric and skew symmetric parts. Clearly the contribution of the skew symmetric part is zero. Notice that $\{\frac{1}{2}[KA_0A + (KA_0A)^t] + A_0B\}(0)$ is positive definite by condition (iii). Thus

$$
\begin{aligned}
(7.17) \quad & \int_0^t \int_{-\infty}^{\infty} \left\langle \frac{\partial^{j+1} u}{\partial x^{j+1}}, KA_0(0)A(0) \frac{\partial^{j+1} u}{\partial x^{j+1}} \right\rangle dx d\tau \\
& = \int_0^t \int_{-\infty}^{\infty} \left\langle \frac{\partial^{j+1} u}{\partial x^{j+1}}, (\frac{1}{2}(KA_0A + (KA_0A)^t) + A_0B)(0) \frac{\partial^{j+1} u}{\partial x^{j+1}} \right\rangle dx d\tau \\
& \quad - \int_0^t \int_{-\infty}^{\infty} \left\langle \frac{\partial^{j+1} u}{\partial x^{j+1}}, A_0(0)B(0) \frac{\partial^{j+1} u}{\partial x^{j+1}} \right\rangle dx d\tau \\
& \geq c \int_0^t \left\| \frac{\partial^{j+1} u}{\partial x^{j+1}} \right\|^2 (\tau) \, d\tau - C \int_0^t \left\| \frac{\partial^{j+1} u^{(2)}}{\partial x^{j+1}} \right\|^2 (\tau) \, d\tau - C N_m(t)^3,
\end{aligned}
$$

where $c > 0$ is a constant, and we have used condition (ii) in the last step. From (7.14) we have for $0 \leq j \leq m-1$,

$$
\begin{aligned}
\left\| \frac{\partial^j h}{\partial x^j} \right\| (t) & \leq \left\| \frac{\partial^j}{\partial x^j} ((A(u) - A(0)) u_x) \right\| (t) + \left\| \frac{\partial^{j+1}}{\partial x^{j+1}} (\tilde{B}(\tilde{u}) u_x^{(2)}) \right\| (t) \\
& \leq C [\|u\|_{H^m}(t) \|u_x\|_{H^{m-1}}(t) + \|u_x^{(2)}\|_{H^m}(t) + \|u_x\|_{H^{m-1}}(t) \|u_x^{(2)}\|_{H^m}(t)].
\end{aligned}
$$

Hence the right-hand side of (7.15) is bounded by

$$
\begin{aligned}
(7.18) \quad & \int_0^t \left\| \frac{\partial^{j+1} u}{\partial x^{j+1}} \right\| \left\| \frac{\partial^j h}{\partial x^j} \right\| (\tau) \, d\tau \leq C \int_0^t \left\| \frac{\partial^{j+1} u}{\partial x^{j+1}} \right\| \|u_x^{(2)}\|_{H^m}(\tau) \, d\tau + C N_m(t)^3 \\
& \leq C \left[\varepsilon \int_0^t \left\| \frac{\partial^{j+1} u}{\partial x^{j+1}} \right\|^2 (\tau) \, d\tau + \frac{1}{4\varepsilon} \int_0^t \|u_x^{(2)}\|_{H^m}^2(\tau) \, d\tau \right] + C N_m(t)^3,
\end{aligned}
$$

where $\varepsilon > 0$ is an arbitrary constant. Choose a small ε. Then (7.15)–(7.18) imply that for $0 \leq t \leq T$, $0 \leq j \leq m-1$,

$$
(7.19) \quad \int_0^t \left\| \frac{\partial^{j+1} u}{\partial x^{j+1}} \right\|^2 (\tau) \, d\tau - C \left\{ \|u\|_{H^m}^2(t) + \int_0^t \|u_x^{(2)}\|_{H^m}^2(\tau) \, d\tau \right\} \leq C \|u_0\|_{H^m}^2 + C N_m(t)^3.
$$

Sum up (7.19) for $0 \leq j \leq m-1$ and multiply the result by a small constant $\varepsilon > 0$. Sum up (7.12) for $1 \leq j \leq m$. The addition of (7.10) and these results gives

$$\|u\|_{H^m}^2(t) + \int_0^t \left(\|u_x\|_{H^{m-1}}^2 + \|u_x^{(2)}\|_{H^m}^2\right)(\tau)\,d\tau \leq C\|u_0\|_{H^m}^2 + CN_m(t)^3, \quad 0 \leq t \leq T.$$

Taking the sup-norm fot t we have $N_m(T)^2 \leq C\|u_0\|_{H^m}^2 + CN_m(T)^3$, which implies $N_m(T) \leq C\|u_0\|_{H^m}$ for small $N_m(T)$. $\square$

8. Systems with Non-Diagonalizable Linearizations

We now prove the main result of this paper, Theorem 2.3. Follow the notations in section 2. Decompose the solution u to (2.1), (2.2) in the right eigenvector directions of $f'(0)$. With λ_i, r_i and l_i being eigenvalues and matrices consisting of right and left eigenvectors of $f'(0)$, given by (2.5), (2.6), we have

$$(8.1) \qquad u(x,t) = \sum_{i=1}^{s} r_i u_i(x,t), \qquad u_i(x,t) = l_i u(x,t), \quad i = 1,\ldots,s.$$

Rewrite (2.1) in terms of the u_i:

$$(8.2) \qquad u_{it} + \lambda_i u_{ix} = l_i B(0) \sum_{j=1}^{s} r_j u_{jxx} + g_{ix} + h_{ix}, \quad i = 1,\ldots,s.$$

where

$$(8.3) \qquad g_i = g_i(u) = \lambda_i u_i - l_i f(u), \qquad h_i = h_i(u) = l_i(B(u) - B(0))u_x.$$

Let the Green's function associated with the linearization of (8.2) be $G(x,t)$. By Duhamel's principle,

$$(8.4) \quad \check{u}(x,t) = \int_{-\infty}^{\infty} G(x-y,t)\check{u}(y,0)\,dy + \int_{0}^{t}\int_{-\infty}^{\infty} G(x-y,t-\tau)[g(y,\tau)+h(y,\tau)]_y\,dyd\tau,$$

where

$$(8.5) \qquad \check{u} = \begin{pmatrix} u_1 \\ \vdots \\ u_s \end{pmatrix}, \qquad g = \begin{pmatrix} g_1 \\ \vdots \\ g_s \end{pmatrix}, \qquad h = \begin{pmatrix} h_1 \\ \vdots \\ h_s \end{pmatrix}.$$

The equations for the diffusion waves are given by (2.11):

$$(8.6) \qquad \theta_{it} + \lambda_i \theta_{ix} + \frac{1}{2}l_i f''(0)(r_i\theta_i, r_i\theta_i)_x = l_i B(0)r_i\theta_{ixx}, \quad i = 1,\ldots,s.$$

Let the Green's function of the linearization of (8.6) be $G^*(x,t)$. We have

$$(8.7) \qquad \theta(x,t) = \int_{-\infty}^{\infty} G^*(x-y,t)\theta(y,0)\,dy + \int_{0}^{t}\int_{-\infty}^{\infty} G^*(x-y,t-\tau)g_y^*(y,\tau)\,dyd\tau,$$

where

(8.8)
$$\theta = \begin{pmatrix} \theta_1 \\ \vdots \\ \theta_s \end{pmatrix}, \qquad g^* = \begin{pmatrix} g_1^* \\ \vdots \\ g_s^* \end{pmatrix},$$

$$g_i^* = -\frac{1}{2} l_i f''(0)(r_i \theta_i, r_i \theta_i), \quad i = 1, \ldots, s.$$

Set

(8.9)
$$\check{v} = \check{u} - \theta = (u_1 - \theta_1, \ldots, u_s - \theta_s)^t \equiv (v_1, \ldots, v_s)^t.$$

Equations (8.4) and (8.7) imply that

(8.10)
$$\begin{aligned}
\check{v}(x,t) &= \int_{-\infty}^{\infty} G^*(x-y,t)\check{v}(y,0)\,dy + \int_{-\infty}^{\infty} (G-G^*)(x-y,t)\check{u}(y,0)\,dy \\
&+ \int_0^t \int_{-\infty}^{\infty} G^*(x-y,t-\tau)(g-g^*+h)_y(y,\tau)\,dyd\tau \\
&+ \int_0^t \int_{-\infty}^{\infty} (G-G^*)(x-y,t-\tau)(g+h)_y(y,\tau)\,dyd\tau.
\end{aligned}$$

As in Section 4, we may assume by Lemma 2.1 that

$$l_i B(0) r_i = \operatorname{diag}(\mu_{i1}, \ldots, \mu_{im_i}), \qquad i = 1, \ldots, s,$$

with $\mu_{ij} > 0$, $j = 1, \ldots, m_i$. Thus the componentwise version of (8.10) takes the form

$$\begin{aligned}
v_{ij}(x,t) &= \int_{-\infty}^{\infty} \frac{1}{\sqrt{4\pi\mu_{ij}t}} e^{-\frac{(x-y-\lambda_i t)^2}{4\mu_{ij}t}} v_{ij}(y,0)\,dy + \int_{-\infty}^{\infty} (G-G^*)_{ij}(x-y,t)\check{u}(y,0)\,dy \\
&+ \int_0^t \int_{-\infty}^{\infty} \frac{1}{\sqrt{4\pi\mu_{ij}(t-\tau)}} e^{-\frac{(x-y-\lambda_i(t-\tau))^2}{4\mu_{ij}(t-\tau)}} (g_{ij} - g_{ij}^* + h_{ij})_y(y,\tau)\,dyd\tau \\
&+ \int_0^t \int_{-\infty}^{\infty} (G-G^*)_{ij}(x-y,t-\tau)(g+h)_y(y,\tau)\,dyd\tau, \\
&\qquad j = 1, \ldots, m_i, \quad i = 1, \ldots, s,
\end{aligned}$$

where v_{ij}, g_{ij}, g_{ij}^* and h_{ij} are, respectively, the jth components of v_i, g_i, g_i^* and h_i, $(G-G^*)_{ij}$ is the corresponding row of $G-G^*$. We have similar formula for higher derivatives. In general, for $l = 0, 1, \ldots,$

$$
\frac{\partial^l}{\partial x^l} v_{ij}(x,t) = \int_{-\infty}^{\infty} \frac{1}{\sqrt{4\pi\mu_{ij}t}} e^{-\frac{(x-y-\lambda_i t)^2}{4\mu_{ij}t}} \frac{\partial^l}{\partial y^l} v_{ij}(y,0)\, dy
$$

$$
+ \int_{-\infty}^{\infty} (G - G^*)_{ij}(x-y,t) \frac{\partial^l}{\partial y^l} \check{u}(y,0)\, dy
$$

(8.11)
$$
+ \int_0^t \int_{-\infty}^{\infty} \frac{1}{\sqrt{4\pi\mu_{ij}(t-\tau)}} e^{-\frac{(x-y-\lambda_i(t-\tau))^2}{4\mu_{ij}(t-\tau)}} \frac{\partial^{l+1}}{\partial y^{l+1}} (g_{ij} - g_{ij}^* + h_{ij})(y,\tau)\, dy d\tau
$$

$$
+ \int_0^t \int_{-\infty}^{\infty} (G - G^*)_{ij}(x-y,t-\tau) \frac{\partial^{l+1}}{\partial y^{l+1}} (g+h)(y,\tau)\, dy d\tau,
$$

$$
j = 1, \ldots, m_i, \quad i = 1, \ldots, s.
$$

Applying Theorem 6.15 to the linearizations of (8.2) and (8.6), we obtain for $l = 0, 1, \ldots,$

(8.12)
$$
\frac{\partial^l}{\partial x^l} (G - G^*)(x,t) = O(1)(t+1)^{-\frac{1}{2}} t^{-\frac{l+1}{2}} \sum_{k=1}^{s} e^{-\frac{(x-\lambda_k t)^2}{Ct}}
$$

$$
+ \sum_{k=1}^{m'} \sum_{k'=0}^{l} e^{-\alpha_k t} \delta^{(l-k')}(x - \beta_k t) Q_{kk'},
$$

where m' is a nonnegative integer not more than the dimension of the null space of $B(0)$, $\alpha_k > 0$ and β_k are constants, and $Q_{kk'}$ are polynomial matrices in t with degrees not more than k', $k = 1, \ldots, m'$, $k' = 0, \ldots, l$.

We now perform a priori estimate through a sequence of lemmas. Adopt the notations in Section 2,

(8.13)
$$
\psi_\alpha(x,t;\lambda) = [(x - \lambda(t+1))^2 + t + 1]^{-\alpha/2},
$$

$$
\phi_\alpha(x,t;\lambda) = [|x - \lambda(t+1)| + t + 1]^{-\alpha/2},
$$

$$
\Phi(x,t) = \sum_{i=1}^{s} [\psi_{3/2}(x,t;\lambda_i) + \phi_1(x,t;\lambda_i)\psi_{3/4}(x,t;\lambda_i)].
$$

Set

(8.14)
$$
M(t) = \sup_{0 \le \tau \le t} \max_{1 \le i \le s} \Big\{ \|v_i(\cdot,\tau)\Phi(\cdot,\tau)^{-1}\|_{L^\infty} + \|v_{ix}(\cdot,\tau)(\tau+1)^{\frac{1}{2}}\Phi(\cdot,\tau)^{-1}\|_{L^\infty}
$$

$$
+ \sum_{l=1}^{2} \Big(\Big\| \frac{\partial^{2l} v_i}{\partial x^{2l}}(\cdot,\tau) \Big\|_{L^\infty} + \Big\| \frac{\partial^{2l+1} v_i}{\partial x^{2l+1}}(\cdot,\tau) \Big\|_{L^\infty} \Big)(\tau+1)^{\frac{3-l}{2}} \Big\}.
$$

Then for $-\infty < x < \infty$, $t \geq 0$, $i = 1, \ldots, s$,

$$|v_i(x,t)| \leq M(t)\Phi(x,t), \qquad |v_{ix}(x,t)| \leq M(t)(t+1)^{-\frac{1}{2}}\Phi(x,t),$$

(8.15)
$$\left|\frac{\partial^2 v_i}{\partial x^2}(x,t)\right|, \quad \left|\frac{\partial^3 v_i}{\partial x^3}(x,t)\right| \leq M(t)(t+1)^{-1},$$

$$\left|\frac{\partial^4 v_i}{\partial x^4}(x,t)\right|, \quad \left|\frac{\partial^5 v_i}{\partial x^5}(x,t)\right| \leq M(t)(t+1)^{-\frac{3}{2}}.$$

Lemma 8.1. *Under the assumptions of Theorem 2.3, we have for $-\infty < x < \infty$, $t \geq 0$, $l = 0$, 1,*

(8.16)
$$\int_{-\infty}^{\infty} \frac{1}{\sqrt{4\pi\mu_{ij}t}} e^{-\frac{(x-y-\lambda_i t)^2}{4\mu_{ij}t}} \frac{\partial^l}{\partial y^l} v_{ij}(y,0)\, dy = O(1)\delta^*(t+1)^{-\frac{l}{2}}\Phi(x,t),$$

where δ^* is the norm of the initial data defined in (2.24).

Proof. By (8.9), (8.1), (2.2), (2.12), (2.20)–(2.24) and (3.1), we have for $1 \leq i \leq s$,

$$v_i(x,0) = u_i(x,0) - \theta_i(x,0) = l_i u_0(x) - \theta_i(x,0), \qquad \|v_i(\cdot,0)\|_{W^{1,1}} = O(1)\delta^*,$$

(8.17)
$$|v_i(x,0)| + |v_{ix}(x,0)| = O(1)\delta^*(1+|x|)^{-5/4}, \qquad \int_{-\infty}^{\infty} v_i(x,0)\, dx = 0,$$

$$\|\eta_i\|_{L^1} = O(1)\delta^*, \qquad |\eta_i(x)| = O(1)\delta^*(1+|x|)^{-1},$$

where

$$\eta_i(x) = \int_{-\infty}^{x} v_i(y,0)\, dy.$$

For $l = 0$, denote the left-hand side of (8.16) as I. For $|x - \lambda_i t| \leq (t+1)^{1/2}$,

$$I = \int_{-\infty}^{\infty} \frac{1}{\sqrt{4\pi\mu_{ij}t}} e^{-\frac{(x-y-\lambda_i t)^2}{4\mu_{ij}t}} \eta'_{ij}(y)\, dy = O(1)\int_{-\infty}^{\infty} t^{-1} e^{-\frac{(x-y-\lambda_i t)^2}{4(\mu_{ij}+\varepsilon)t}} |\eta_{ij}(y)|\, dy$$

$$= O(1)\delta^*(t+1)^{-1} = O(1)\delta^*\psi_{3/2}(x,t;\lambda_i),$$

where η_{ij} is the jth component of η_i, and ε a small positive constant. We have assumed $t \geq 1$ in the above estimate. The case $t \leq 1$ is clear. For $(t+1)^{1/2} \leq |x - \lambda_i t| \leq t+1$, say, $x - \lambda_i t > 0$, then

(8.18)
$$I = O(1)\int_{-\infty}^{\infty} t^{-1} e^{-\frac{(x-y-\lambda_i t)^2}{4(\mu_{ij}+\varepsilon)t}} |\eta_{ij}(y)|\, dy$$

$$= O(1)\int_{-\infty}^{(x-\lambda_i t)/2} t^{-1} e^{-\frac{(x-\lambda_i t)^2}{Ct}} |\eta_{ij}(y)|\, dy$$

$$+ O(1)\int_{(x-\lambda_i t)/2}^{\infty} t^{-1} e^{-\frac{(x-y-\lambda_i t)^2}{Ct}} \delta^*(1+y)^{-1}\, dy$$

$$= O(1)\delta^* t^{-1} e^{-\frac{(x-\lambda_i t)^2}{Ct}} + O(1)\delta^* t^{-\frac{1}{2}}(1+|x-\lambda_i t|)^{-1}$$

$$= O(1)\delta^*(t+1)^{-\frac{1}{2}}|x-\lambda_i t|^{-1} = O(1)\delta^*\psi_{3/2}(x,t;\lambda_i).$$

The case that $x - \lambda_i t < 0$ is similar. For $|x - \lambda_i t| \geq t + 1$, again consider $x - \lambda_i t > 0$. Then

(8.19)
$$
\begin{aligned}
I &= O(1) \int_{-\infty}^{(x-\lambda_i t)/2} t^{-\frac{1}{2}} e^{-\frac{(x-\lambda_i t)^2}{Ct}} |v_{ij}(y,0)| \, dy \\
&\quad + O(1) \int_{(x-\lambda_i t)/2}^{\infty} t^{-\frac{1}{2}} e^{-\frac{(x-y-\lambda_i t)^2}{Ct}} \delta^* (1+y)^{-\frac{5}{4}} \, dy \\
&= O(1)\delta^* t^{-\frac{1}{2}} e^{-\frac{(x-\lambda_i t)^2}{Ct}} + O(1)\delta^* |x - \lambda_i t|^{-\frac{5}{4}} \\
&= O(1)\delta^* |x - \lambda_i t|^{-\frac{5}{4}} = O(1)\delta^* \phi_1(x,t;\lambda_i)\psi_{3/4}(x,t;\lambda_i).
\end{aligned}
$$

Similarly the estimate holds for $x - \lambda_i t < 0$. We thus have proved (8.16) for $l = 0$. For $l = 1$, the case that $t \geq 1$ is clear by integrating the left-hand side by parts; while the case that $t \leq 1$ is similar to the case when $l = 0$. $\qquad\square$

Lemma 8.2. *Under the assumptions of Theorem 2.3, we have for* $-\infty < x < \infty$, $t \geq 0$,

(8.20)
$$
\int_{-\infty}^{\infty} \frac{1}{\sqrt{4\pi\mu_{ij}t}} e^{-\frac{(x-y-\lambda_i t)^2}{4\mu_{ij}t}} \frac{\partial^l}{\partial y^l} v_{ij}(y,0) \, dy = O(1)\delta^*(t+1)^{-\frac{l+1}{2}},
$$

where $2 \leq l \leq 5$ *is an integer.*

Proof. Denote the left-hand side of (8.20) as I. For $t \geq 1$, we perform integration by parts,

$$
I = O(1) \int_{-\infty}^{\infty} t^{-\frac{l+1}{2}} e^{-\frac{(x-y-\lambda_i t)^2}{Ct}} |v_{ij}(y,0)| \, dy = O(1)(t+1)^{-\frac{l+1}{2}} \|v_{ij}(\cdot,0)\|_{L^1}.
$$

Equation (8.20) follows by (8.17). For $t \leq 1$,

$$
I = O(1) \left\| \frac{\partial^l}{\partial x^l} v_{ij}(\cdot,0) \right\|_{L^\infty} = O(1)\|v_{ij}(\cdot,0)\|_{H^{l+1}} = O(1)\delta^*(t+1)^{-\frac{l+1}{2}}
$$

by (2.24). $\qquad\square$

Lemma 8.3. *Under the assumptions of Theorem 2.3, we have for* $-\infty < x < \infty$, $t \geq 0$, $l = 0, 1$,

(8.21)
$$
\int_{-\infty}^{\infty} (G - G^*)_{ij}(x - y, t) \frac{\partial^l}{\partial y^l} \breve{u}(y,0) \, dy = O(1)\delta^*(t+1)^{-\frac{1}{4}-\frac{l}{2}} \Phi(x,t).
$$

Proof. By (8.12), (8.5), (8.1) and (2.24),

(8.22)
$$
\begin{aligned}
I &\equiv \int_{-\infty}^{\infty} (G - G^*)_{ij}(x - y, t) \breve{u}(y,0) \, dy \\
&= \int_{-\infty}^{\infty} O(1)(t+1)^{-\frac{1}{2}} t^{-\frac{1}{2}} \sum_{k=1}^{s} e^{-\frac{(x-y-\lambda_k t)^2}{Ct}} |u_0(y)| \, dy + \sum_{k=1}^{m'} O(1) e^{-t/C} |u_0(x - \beta_k t)| \\
&= O(1) \sum_{k=1}^{s} \int_{-\infty}^{\infty} (t+1)^{-\frac{1}{2}} t^{-\frac{1}{2}} e^{-\frac{(x-y-\lambda_k t)^2}{Ct}} |u_0(y)| \, dy + O(1)\delta^* e^{-t/C}(1 + |x|)^{-\frac{5}{4}}.
\end{aligned}
$$

Set

(8.23)
$$I_k = \int_{-\infty}^{\infty} (t+1)^{-\frac{1}{2}} t^{-\frac{1}{2}} e^{-\frac{(x-y-\lambda_k t)^2}{Ct}} |u_0(y)| \, dy.$$

By (2.24), for $|x - \lambda_k t| \le (t+1)^{\frac{1}{2}}$,

$$I_k = O(1)\delta^*(t+1)^{-1} = O(1)\delta^*(t+1)^{-\frac{1}{4}} \psi_{3/2}(x,t;\lambda_k).$$

Similar to (8.18), for $(t+1)^{1/2} \le |x - \lambda_k t| \le t+1$, we have by (2.24)

$$I_k = O(1)\delta^*(t+1)^{-\frac{1}{2}} t^{-\frac{1}{2}} e^{-\frac{(x-\lambda_k t)^2}{Ct}} + O(1)\delta^*(t+1)^{-\frac{1}{2}}(1 + |x - \lambda_k t|)^{-\frac{5}{4}}$$
$$= O(1)\delta^*(t+1)^{-\frac{1}{4}} \psi_{3/2}(x,t;\lambda_k).$$

And similar to (8.19), for $|x - \lambda_k t| \ge t+1$, we have

$$I_k = O(1)\delta^*(t+1)^{-\frac{1}{2}} \phi_1(x,t;\lambda_k)\psi_{3/4}(x,t;\lambda_k).$$

Together with (8.22), (8.23) and (8.13),

$$I = O(1)\delta^*(t+1)^{-\frac{1}{4}} \Phi(x,t).$$

Therefore, (8.21) is proved for $l = 0$. For $l = 1$, (8.21) follows by integration by parts.

$\square$

Lemma 8.4. *Under the assumptions of Theorem 2.3, we have for $-\infty < x < \infty$, $t \ge 0$, $l = 2, \ldots, 5$,*

(8.24)
$$\int_{-\infty}^{\infty} (G - G^*)_{ij}(x - y, t) \frac{\partial^l}{\partial y^l} \check{u}(y,0) \, dy = O(1)\delta^*(t+1)^{-1-\frac{l}{2}}.$$

Proof. Similar to the proof of (8.20), (8.24) is clear by integration by parts and (8.12).

$\square$

From (8.3), (8.8) and (8.9), we have the following expressions:

(8.25)
$$g_i - g_i^* = \lambda_i u_i - l_i f(u) + \frac{1}{2} l_i f''(0)(r_i \theta_i, r_i \theta_i)$$
$$= -\frac{1}{2} l_i \sum_{j \ne i} f''(0)(r_j \theta_j, r_j \theta_j) - l_i \sum_{j=1}^{s} f''(0)(r_j \theta_j, r_j v_j)$$
$$+ O(1) \sum_{j \ne k} |\theta_j|(|v_k| + |\theta_k|) + O(1) \sum_{j=1}^{s} |v_j|^2 + O(1) \sum_{j=1}^{s} |\theta_j|^3,$$

(8.26)
$$(g_i - g_i^*)_x = -\frac{1}{2} l_i \sum_{j \neq i} f''(0)(r_j \theta_j, r_j \theta_j)_x + O(1) \sum_{j,k=1}^{s} (|\theta_j||v_{kx}|$$

$$+ |\theta_{jx}||v_k| + |\theta_j|^2|\theta_{kx}| + |v_j||v_{kx}|) + O(1) \sum_{j \neq k} |\theta_{jx}||\theta_k|,$$

(8.27)
$$\left\| \frac{\partial^l}{\partial x^l} g_i(\cdot, t) \right\|_{L^\infty} = O(1) \left[\|u(\cdot,t)\|_{L^\infty} \left\| \frac{\partial^l u}{\partial x^l}(\cdot, t) \right\|_{L^\infty} + \|u_x(\cdot, t)\|_{W^{l-2,\infty}}^2 \right],$$

$$l \geq 2.$$

Let θ_α be defined by (3.8):

(8.28)
$$\theta_\alpha(x, t; \lambda, \mu) = (t+1)^{-\frac{\alpha}{2}} e^{-\frac{(x-\lambda(t+1))^2}{\mu(t+1)}},$$

where $\alpha > 0$, $\mu > 0$ and λ are constants.

Lemma 8.5. *Under the assumptions of Theorem 2.3, we have for* $-\infty < x < \infty$, $t \geq 0$, $l = 0, 1$,

(8.29)
$$\int_0^t \int_{-\infty}^\infty \frac{1}{\sqrt{4\pi\mu_{ij}(t-\tau)}} e^{-\frac{(x-y-\lambda_i(t-\tau))^2}{4\mu_{ij}(t-\tau)}} \frac{\partial^{l+1}}{\partial y^{l+1}} (g_{ij} - g_{ij}^*)(y, \tau) \, dy d\tau$$

$$= O(1)(\delta^* + M(t))^2 (t+1)^{-\frac{l}{2}} \Phi(x, t).$$

Proof. Denote the left-hand side of (8.29) as I. By (8.25), (8.26), (3.1), (8.15), (8.13) and integration by parts, we have for $l = 0, 1$,

$$I = \int_0^t \int_{-\infty}^\infty \frac{1}{\sqrt{4\pi\mu_{ij}(t-\tau)}} e^{-\frac{(x-y-\lambda_i(t-\tau))^2}{4\mu_{ij}(t-\tau)}}$$

$$\cdot \frac{\partial^{l+1}}{\partial y^{l+1}} \left[-\frac{1}{2} l_{ij} \sum_{k \neq i} f''(0)(r_k \theta_k, r_k \theta_k)(y, \tau) \right] dy d\tau$$

(8.30)
$$+ O(1) \int_0^{\frac{t}{2}} \int_{-\infty}^\infty (t-\tau)^{-\frac{1}{2}-1} e^{-\frac{(x-y-\lambda_i(t-\tau))^2}{4(\mu_{ij}+\varepsilon)(t-\tau)}} \Big[\delta^*(\delta^* + M(\tau)) \sum_k \theta_{5/2}(y, \tau; \lambda_k, \mu)$$

$$+ M(\tau)^2 \sum_k (\tau+1)^{-\frac{3}{4}} \psi_{3/2}(y, \tau; \lambda_k) \Big] dy d\tau$$

$$+ O(1) \int_{\frac{t}{2}}^t \int_{-\infty}^\infty (t-\tau)^{-1} e^{-\frac{(x-y-\lambda_i(t-\tau))^2}{4(\mu_{ij}+\varepsilon)(t-\tau)}} \Big[\delta^*(\delta^* + M(\tau)) \sum_k \theta_{5/2+l}(y, \tau; \lambda_k, \mu)$$

$$+ M(\tau)^2 \sum_k (\tau+1)^{-\frac{3}{4}-\frac{l}{2}} \psi_{3/2}(y, \tau; \lambda_k) \Big] dy d\tau$$

$$\equiv I_1 + I_2 + I_3,$$

where ε, $\mu > 0$ are constants. We see that the first integral I_1 in (8.30) is II in (4.14) when $l = 0$, and $\widetilde{II}$ in (4.16) when $l = 1$. Hence by (4.27) and (4.33),

$$(8.31) \qquad I_1 = O(1)\delta^{*2}(t+1)^{-\frac{l}{2}} \sum_{k=1}^{s} \psi_{3/2}(x,t;\lambda_k).$$

To estimate the second integral in (8.30), we use Lemmas 3.2, 3.3, 3.5 and 3.6:

$$
\begin{aligned}
(8.32) \qquad I_2 &= O(1)(\delta^* + M(t))^2 t^{-\frac{l}{2}} \int_0^{t/2} \int_{-\infty}^{\infty} (t-\tau)^{-1} e^{-\frac{(x-y-\lambda_i(t-\tau))^2}{4(\mu_{ij}+\varepsilon)(t-\tau)}} \\
&\quad \cdot \sum_k [\theta_{5/2}(y,\tau;\lambda_k,\mu) + (\tau+1)^{-\frac{3}{4}} \psi_{3/2}(y,\tau;\lambda_k)]\, dy d\tau \\
&= O(1)(\delta^* + M(t))^2 t^{-\frac{l}{2}} \sum_k [\psi_{3/2}(x,t;\lambda_k) + \phi_1(x,t;\lambda_k)\psi_{3/4}(x,t;\lambda_k)],
\end{aligned}
$$

where we have used (3.24) and the fact that for $K \geq |\lambda_i - \lambda_k|$,

$$
\begin{aligned}
(8.33) \qquad &|x - \lambda_i(t+1)|^{-\frac{3}{4}} |x - \lambda_k(t+1)|^{-\frac{1}{2}} \operatorname{char}\{\min(\lambda_i,\lambda_k)(t+1) + K\sqrt{t+1} \leq x \\
&\qquad \leq \max(\lambda_i,\lambda_k)(t+1) - K\sqrt{t+1}\} \\
&= O(1)[\phi_1(x,t;\lambda_i)\psi_{3/4}(x,t;\lambda_i) + \phi_1(x,t;\lambda_k)\psi_{3/4}(x,t;\lambda_k)].
\end{aligned}
$$

The third integral in (8.30) is estimated in the same way,

$$(8.34) \qquad I_3 = O(1)(\delta^* + M(t))^2 (t+1)^{-\frac{l}{2}} \Phi(x,t).$$

Hence (8.29) is proved by (8.30)-(8.32) and (8.34). $\qquad\qquad\qquad\qquad\qquad\square$

Lemma 8.6. *Under the assumptions of Theorem 2.3, we have for $-\infty < x < \infty$, $t \geq 0$,*

$$
\begin{aligned}
(8.35) \quad \int_0^t \int_{-\infty}^{\infty} \frac{1}{\sqrt{4\pi\mu_{ij}(t-\tau)}} &e^{-\frac{(x-y-\lambda_i(t-\tau))^2}{4\mu_{ij}(t-\tau)}} \frac{\partial^{l+1}}{\partial y^{l+1}}(g_{ij} - g_{ij}^*)(y,\tau)\, dy d\tau \\
&= O(1)(\delta^* + M(t))^2 \begin{cases} (t+1)^{-1}, & \text{if } l = 2, 3 \\ (t+1)^{-\frac{1}{2}}, & \text{if } l = 4, 5 \end{cases}.
\end{aligned}
$$

Proof. Denote the left-hand side of (8.35) as I. By integration by parts,

$$
\begin{aligned}
I = O(1) \int_0^{t/2} (t-\tau)^{-\frac{1}{2}-1} \|(g_{ij} - g_{ij}^*)(\cdot,\tau)\|_{L^1}\, d\tau \\
+ O(1) \int_{t/2}^t (t-\tau)^{-\frac{1}{2}} \left\| \frac{\partial^l}{\partial y^l}(g_{ij} - g_{ij}^*)(\cdot,\tau) \right\|_{L^\infty}\, d\tau.
\end{aligned}
$$

By (8.25), (3.1), (8.15) and (8.13) it is clear that

$$\|(g_{ij} - g_{ij}^*)(\cdot, \tau)\|_{L^1} = O(1) \int_{-\infty}^{\infty} \Big[\sum_k (\delta^* + M(\tau))(\tau+1)^{-\frac{1}{2}} |\theta_k(y, \tau)| + (M(\tau)\Phi(y, \tau))^2 \Big] \, dy$$

$$= O(1)(\delta^* + M(\tau))^2 (\tau+1)^{-\frac{1}{2}}.$$

Also by (8.27), (8.1), (8.9), (8.8) and Lemma 2.2,

$$\left\| \frac{\partial^l}{\partial y^l} (g_{ij} - g_{ij}^*)(\cdot, \tau) \right\|_{L^\infty} = O(1) \Bigg[\Big(\sum_k \|\theta_k + v_k\|_{L^\infty} \Big) \Big(\sum_k \left\| \frac{\partial^l}{\partial y^l}(\theta_k + v_k) \right\|_{L^\infty} \Big)$$

$$+ \Big(\sum_k \|\theta_{ky} + v_{ky}\|_{W^{l-2,\infty}} \Big)^2 + \left\| \frac{\partial^l}{\partial y^l} f''(0)_{(r_i\theta_i, r_i\theta_i)} \right\|_{L^\infty} \Bigg]$$

$$= O(1)(\delta^* + M(\tau))^2 \begin{cases} (\tau+1)^{-3/2}, & \text{if } l = 2, 3 \\ (\tau+1)^{-1}, & \text{if } l = 4, 5 \end{cases}.$$

Thus (8.35) is straightforward.   □

By (8.3) we have

$$h_{ij} = l_{ij}(B(u) - B(0))u_x = b_{ij}(u, u_x) + O(1)|u|^2 |u_x|$$

(8.36)
$$= \sum_{k=1}^{s} b_{ij}(r_k\theta_k, r_k\theta_{kx}) + O(1) \sum_{k \neq k'} |\theta_k||\theta_{k'x}|$$

$$+ O(1) \sum_{k,k'} [|v_k||\theta_{k'x}| + |\theta_k||v_{k'x}| + |v_k||v_{k'x}| + |\theta_k|^2|\theta_{k'x}|],$$

where b_{ij} is a bilinear form. Similarly,

(8.37)
$$h_{ijx} = O(1)(|u||u_{xx}| + |u_x|^2)$$

$$= O(1) \Bigg[\sum_{k,k'} (|\theta_k||\theta_{k'xx}| + |v_k||\theta_{k'xx}| + |\theta_k||v_{k'xx}| + |v_k||v_{k'xx}|)$$

$$+ \sum_k (|\theta_{kx}|^2 + |v_{kx}|^2) \Bigg],$$

(8.38)
$$\left\| \frac{\partial^l}{\partial x^l} h_{ij}(\cdot, t) \right\|_{L^\infty} = O(1) \Bigg[\|u(\cdot, t)\|_{L^\infty} \left\| \frac{\partial^{l+1}}{\partial x^{l+1}} u(\cdot, t) \right\|_{L^\infty}$$

$$+ \|u_x(\cdot, t)\|_{L^\infty} \left\| \frac{\partial^l}{\partial x^l} u(\cdot, t) \right\|_{L^\infty} + \|u_x(\cdot, t)\|_{W^{l-2,\infty}}^2 \Bigg], \quad l \geq 2.$$

Lemma 8.7. *Under the assumptions of Theorem 2.3, we have for* $-\infty < x < \infty$, $t \geq 0$,

$$(8.39) \quad \int_0^t \int_{-\infty}^{\infty} \frac{1}{\sqrt{4\pi\mu_{ij}(t-\tau)}} e^{-\frac{(x-y-\lambda_i(t-\tau))^2}{4\mu_{ij}(t-\tau)}} \frac{\partial^{l+1}}{\partial y^{l+1}} h_{ij}(y,\tau)\, dy d\tau$$

$$= O(1)(\delta^* + M(t))^2 \begin{cases} (t+1)^{-\frac{1}{2}-\frac{1}{4}+\alpha}\Phi(x,t), & \text{if } l = 0, 1 \\ (t+1)^{-1}, & \text{if } l = 2, 3, \\ (t+1)^{-\frac{1}{2}}, & \text{if } l = 4, 5 \end{cases}$$

where $\alpha > 0$ *is arbitrarily small. Moreover, if all the* λ_k, $1 \leq k \leq s$, *are simple,* $\alpha = 0$.

Proof. Denote the left-hand side of (8.39) as I. For $l = 0, 1$, we have by (8.36), (8.37), (3.1), (8.15) and (8.13),

$$(8.40)$$
$$I = \int_0^t \int_{-\infty}^{\infty} \frac{1}{\sqrt{4\pi\mu_{ij}(t-\tau)}} e^{-\frac{(x-y-\lambda_i(t-\tau))^2}{4\mu_{ij}(t-\tau)}} \sum_k \frac{\partial^{l+1}}{\partial y^{l+1}} b_{ij}(r_k\theta_k, r_k\theta_{kx})(y,\tau)\, dy d\tau$$
$$+ O(1) \int_0^{t-1} \int_{-\infty}^{\infty} (t-\tau)^{-\frac{l}{2}-1} e^{-\frac{(x-y-\lambda_i(t-\tau))^2}{4(\mu_{ij}+\varepsilon)(t-\tau)}} \left[\delta^*(\delta^* + M(\tau)) \sum_k \theta_{\frac{7}{2}}(y,\tau;\lambda_k,\mu) \right.$$
$$\left. + M(\tau)^2(\tau+1)^{-\frac{5}{4}} \sum_k \psi_{3/2}(y,\tau;\lambda_k) \right] dy d\tau$$
$$+ O(1) \int_{t-1}^t \int_{-\infty}^{\infty} (t-\tau)^{-1} e^{-\frac{(x-y-\lambda_i(t-\tau))^2}{4(\mu_{ij}+\varepsilon)(t-\tau)}} \left[\delta^*(\delta^* + M(\tau)) \sum_k \theta_3(y,\tau;\lambda_k,\mu) \right.$$
$$\left. + M(\tau)^2(\tau+1)^{-1} \sum_k (\psi_{3/2}(y,\tau;\lambda_k) + \phi_1(y,\tau;\lambda_k)\psi_{3/4}(y,\tau;\lambda_k)) \right] dy d\tau$$
$$\equiv I_1 + I_2 + I_3,$$

where ε, $\mu > 0$ are constants. To estimate I_1, we use Lemma 3.2 and integrate by parts for $k = i$, and use Lemma 3.4 for $k \neq i$. Thus

$$I_1 = O(1)\delta^{*2} \left[\theta_{2+l}(x,t;\lambda_i,\mu) \log(t+2) + \sum_k (t+1)^{-\frac{l}{2}-\frac{1}{4}+\alpha} \psi_{3/2}(x,t;\lambda_k) \right]$$
$$= O(1)\delta^{*2}(t+1)^{-\frac{l}{2}-\frac{1}{4}+\alpha} \sum_k \psi_{3/2}(x,t;\lambda_k).$$

Then we apply Lemmas 3.2, 3.3, 3.5 and 3.6 to I_2:

$$I_2 = O(1)(\delta^* + M(t))^2 \Bigg\{ (t+1)^{-\frac{l}{2}} \int_0^{t/2} \int_{-\infty}^{\infty} (t-\tau)^{-1} e^{-\frac{(x-y-\lambda_i(t-\tau))^2}{4(\mu_{ij}+\varepsilon)(t-\tau)}}$$

$$\cdot \sum_k [\theta_{7/2}(y,\tau;\lambda_k,\mu) + (\tau+1)^{-\frac{5}{4}}\psi_{3/2}(y,t;\lambda_k)]\, dy d\tau$$

$$+ (t+1)^{-1} \int_{t/2}^{t} \int_{-\infty}^{\infty} (t-\tau)^{-1}(t+1-\tau)^{-\frac{l}{2}} e^{-\frac{(x-y-\lambda_i(t-\tau))^2}{4(\mu_{ij}+\varepsilon)(t-\tau)}}$$

$$\cdot \sum_k [\theta_{3/2}(y,\tau;\lambda_k,\mu) + (\tau+1)^{-\frac{1}{4}}\psi_{3/2}(y,\tau;\lambda_k)]\, dy d\tau \Bigg\}$$

$$= O(1)(\delta^* + M(t))^2 (t+1)^{-\frac{l}{2}-\frac{1}{4}} \sum_k \psi_{3/2}(x,t;\lambda_k).$$

To estimate I_3, notice that $(t+1-\tau) = O(1)$ for $t-1 \le \tau \le t$. Thus

$$I_3 = O(1)(\delta^* + M(t))^2 (t+1)^{-\frac{3}{4}} \int_{t-1}^{t} \int_{-\infty}^{\infty} (t-\tau)^{-1}(t+1-\tau)^{-1} e^{-\frac{(x-y-\lambda_i(t-\tau))^2}{4(\mu_{ij}+\varepsilon)(t-\tau)}}$$

$$\cdot \sum_k [\theta_{3/2}(y,\tau;\lambda_k,\mu) + (\tau+1)^{-\frac{1}{4}}(\psi_{3/2}(y,t;\lambda_k) + \phi_1(y,\tau;\lambda_k)\psi_{3/4}(y,\tau,\lambda_k))]\, dy d\tau$$

$$= O(1)(\delta^* + M(t))^2 (t+1)^{-\frac{3}{4}} \sum_k (\psi_{3/2}(x,t;\lambda_k) + \phi_1(x,t;\lambda_k)\psi_{3/4}(y,\tau;\lambda_k))$$

by Lemmas 3.2, 3.3, 3.5–3.8. We have proved (8.39) for $l = 0$, 1. We see that $\alpha > 0$ appears only in I_1. If all the λ_k are simple, then the θ_k are scalars, and

$$b_{ij}(r_k\theta_k, r_k\theta_{kx}) = b_{ij}(r_k, r_k)\left(\frac{\theta_k^2}{2}\right)_x.$$

We can avoid α in this case. For $2 \le l \le 5$, we have

$$I = O(1) \int_0^{t-1} \int_{-\infty}^{\infty} (t-\tau)^{-\frac{l}{2}-1} e^{-\frac{(x-y-\lambda_i(t-\tau))^2}{4(\mu_{ij}+\varepsilon)(t-\tau)}} h_{ij}(y,\tau)\, dy d\tau$$

$$+ O(1) \int_{t-1}^{t} \int_{-\infty}^{\infty} (t-\tau)^{-1} e^{-\frac{(x-y-\lambda_i(t-\tau))^2}{4(\mu_{ij}+\varepsilon)(t-\tau)}} \frac{\partial^l}{\partial y^l} h_{ij}(y,\tau)\, dy d\tau$$

$$= O(1) \int_0^{t-1} (t-\tau)^{-\frac{l}{2}-\frac{1}{2}} \|h_{ij}(\cdot,\tau)\|_{L^\infty}\, d\tau + O(1) \int_{t-1}^{t} (t-\tau)^{-\frac{1}{2}} \left\|\frac{\partial^l}{\partial y^l} h_{ij}(\cdot,\tau)\right\|_{L^\infty}\, d\tau$$

$$\equiv I_4 + I_5.$$

From (8.36), (8.15),

$$I_4 = O(1) \int_0^{t-1} (t-\tau)^{-\frac{l}{2}-\frac{1}{2}} (\delta^* + M(\tau))^2 (\tau+1)^{-\frac{3}{2}}\, d\tau$$

$$= O(1)(\delta^* + M(t))^2 \left[t^{-\frac{l}{2}-\frac{1}{2}} \int_0^{t/2} (\tau+1)^{-\frac{3}{2}}\, d\tau + (t+1)^{-\frac{3}{2}} \int_{t/2}^{t-1} (t-\tau)^{-\frac{l}{2}-\frac{1}{2}}\, d\tau \right]$$

$$= O(1)(\delta^* + M(t))^2 [(t+1)^{-\frac{l}{2}-\frac{1}{2}} + (t+1)^{-\frac{3}{2}}],$$

which is bounded by the right-hand side of (8.39). By a standard interpolation formula and Theorem 7.1, we have for $t \geq 0$,

$$(8.41) \qquad \|u(\cdot,t)\|_{W^{7,\infty}} = O(1)\|u(\cdot,t)\|_{H^8} = O(1)\delta^*.$$

Together with (8.38) and (8.15),

$$\left\|\frac{\partial^l}{\partial x^l} h_{ij}(\cdot,t)\right\|_{L^\infty} = O(1)(\delta^* + M(t))^2 \begin{cases} (t+1)^{-1}, & \text{if } l = 2,\, 3 \\ (t+1)^{-\frac{1}{2}}, & \text{if } l = 4,\, 5 \end{cases}.$$

Thus

$$I_5 = O(1)(\delta^* + M(t))^2 \int_{t-1}^t (t-\tau)^{-\frac{1}{2}}\, d\tau \begin{cases} (t+1)^{-1}, & \text{if } l = 2,\, 3 \\ (t+1)^{-\frac{1}{2}}, & \text{if } l = 4,\, 5 \end{cases},$$

which is also bounded by the right-hand side of (8.39). $\qquad\square$

Lemma 8.8. *Under the assumptions of Theorem 2.3, we have for $-\infty < x < \infty$, $t \geq 0$,*

$$(8.42) \qquad \int_0^t \int_{-\infty}^\infty (G - G^*)_{ij}(x - y, t - \tau)\frac{\partial^{l+1}}{\partial y^{l+1}}(g + h)(y, \tau)\, dy d\tau$$

$$= O(1)(\delta^* + M(t))^2 \begin{cases} (t+1)^{-\frac{l}{2}-\frac{1}{4}}\Phi(x,t), & \text{if } l = 0,\, 1 \\ (t+1)^{-1}, & \text{if } l = 2,\, 3 \\ (t+1)^{-\frac{1}{2}}, & \text{if } l = 4,\, 5 \end{cases}.$$

Proof. Denote the left-hand side of (8.42) as I. Consider $l = 0,\, 1$. We have

$$I = I_1 + I_2 + I_3,$$

where

$$I_1 = \int_0^t \int_{-\infty}^\infty (G - G^*)_{ij}(x - y, t - \tau)\frac{\partial^{l+1}}{\partial y^{l+1}} \sum_k \left(-\frac{1}{2}Lf''(0)(r_k\theta_k, r_k\theta_k)\right.$$
$$\left. + b(r_k\theta_k, r_k\theta_{kx})\right)(y, \tau)\, dy d\tau,$$

$$I_2 = \int_0^t \int_{-\infty}^\infty (G - G^*)_{ij}(x - y, t - \tau)\frac{\partial^{l+1}}{\partial y^{l+1}}\left(g + \sum_k \frac{1}{2}Lf''(0)(r_k\theta_k, r_k\theta_k)\right)(y, \tau)\, dy d\tau,$$

$$I_3 = \int_0^t \int_{-\infty}^\infty (G - G^*)_{ij}(x - y, t - \tau)\frac{\partial^{l+1}}{\partial y^{l+1}}\left(h - \sum_k b(r_k\theta_k, r_k\theta_{kx})\right)(y, \tau)\, dy d\tau,$$

and $b(w_1, w_2)$ is the n-vector valued bilinear form of w_1, $w_2 \in \mathbb{R}^n$ which has $b_{ij}(w_1, w_2)$ as the jth component of its ith block. To estimate I_1, we integrate by parts with respect to

x for $0 \leq \tau \leq t/2$ and use (8.12):

$$I_1 = \int_0^{t/2} \int_{-\infty}^{\infty} O(1)(t-\tau)^{-1-\frac{l+1}{2}} \sum_{k,k'=1}^{s} e^{-\frac{(x-y-\lambda_k(t-\tau))^2}{C(t-\tau)}} \delta^{*2} \theta_2(y,\tau;\lambda_{k'},\mu)\,dy d\tau$$

$$+ \int_0^{t/2} O(1)\delta^{*2} e^{-(t-\tau)/C} \sum_{k=1}^{m'} \sum_{k'=1}^{s} \theta_2(x-\beta_k(t-\tau),\tau;\lambda_{k'},\mu)\,d\tau$$

$$+ \int_{t/2}^{t} \int_{-\infty}^{\infty} O(1)(t-\tau)^{-\frac{1}{2}}(t+1-\tau)^{-\frac{1}{2}} \sum_{k,k'=1}^{s} e^{-\frac{(x-y-\lambda_k(t-\tau))^2}{C(t-\tau)}} \delta^{*2} \theta_{3+l}(y,\tau;\lambda_{k'},\mu)\,dy d\tau$$

$$+ \int_{t/2}^{t} O(1)\delta^{*2} e^{-(t-\tau)/C} \sum_{k=1}^{m'} \sum_{k'=1}^{s} \theta_{3+l}(x-\beta_k(t-\tau),\tau;\lambda_{k'},\mu)\,d\tau$$

$$=O(1)\delta^{*2}(t+1)^{-\frac{l+1}{2}} \Bigg[\sum_{k,k'=1}^{s} \int_0^t \int_{-\infty}^{\infty} (t-\tau)^{-1} e^{-\frac{(x-y-\lambda_k(t-\tau))^2}{C(t-\tau)}} \theta_2(y,\tau;\lambda_{k'},\mu)\,dy d\tau$$

$$+ \sum_{k=1}^{m'} \sum_{k'=1}^{s} \int_0^t e^{-(t-\tau)/C} \theta_2(x-\beta_k(t-\tau),\tau;\lambda_{k'},\mu)\,d\tau \Bigg],$$

where $\mu > 0$ is a constant. Applying Lemmas 3.2, 3.3 to the first term, Lemma 3.9 to the second term, and using (8.33), (3.24), we see that I_1 is bounded by the right-hand side of (8.42). By (8.3), (8.9), (3.1), (8.15), we have

$$\frac{\partial^2 g}{\partial x^2} = O(1)(|u||u_{xx}| + |u_x|^2)$$

$$= O(1)(\delta^* + M(t))^2 \Bigg[\sum_{k=1}^{s} \theta_3(x,t;\lambda_k,\mu) + (t+1)^{-1}\Phi(x,t) \Bigg],$$

$$\frac{\partial^2 h}{\partial x^2} = O(1)(|u||u_{xxx}| + |u_x||u_{xx}| + |u_x|^3)$$

$$= O(1)(\delta^* + M(t))^2 \Bigg[\sum_{k=1}^{s} \theta_3(x,t;\lambda_k,\mu) + (t+1)^{-1}\Phi(x,t) \Bigg].$$

Together with (8.26), (8.8), (8.36), (8.37), and making use of (8.12), we estimate I_2 and I_3 in a similar way as I_1:

$$I_2 = \int_0^t \int_{-\infty}^{\infty} O(1)(t-\tau)^{-\frac{1}{2}-\frac{l}{2}}(t+1-\tau)^{-\frac{1}{2}} \sum_{k,k'=1}^{s} e^{-\frac{(x-y-\lambda_k(t-\tau))^2}{C(t-\tau)}} (\delta^* + M(\tau))^2$$

$$\cdot [\theta_{7/2}(y,\tau;\lambda_{k'},\mu) + (\tau+1)^{-\frac{5}{4}}\psi_{3/2}(y,\tau;\lambda_{k'})]\,dy d\tau$$

$$+ \int_0^t O(1)e^{-(t-\tau)/C} \sum_{k=1}^{m'} (\delta^* + M(\tau))^2 \Bigg[\sum_{k'=1}^{s} \theta_3(x-\beta_k(t-\tau),\tau;\lambda_{k'},\mu)$$

$$+ (\tau + 1)^{-1} \Phi(x - \beta_k(t - \tau), \tau) \bigg] d\tau,$$

$$I_3 = \int_0^{t-1} \int_{-\infty}^{\infty} O(1)(t - \tau)^{-1-\frac{l+1}{2}} \sum_{k,k'=1}^{s} e^{-\frac{(x-y-\lambda_k(t-\tau))^2}{C(t-\tau)}} (\delta^* + M(\tau))^2$$

$$\cdot \left[\theta_{7/2}(y, \tau; \lambda_{k'}, \mu) + (\tau + 1)^{-\frac{5}{4}} \psi_{3/2}(y, \tau; \lambda_{k'}) \right] dy d\tau$$

$$+ \int_0^t O(1) e^{-(t-\tau)/C} \sum_{k=1}^{m'} (\delta^* + M(\tau))^2 \left[\sum_{k'=1}^{s} \theta_3(x - \beta_k(t - \tau), \tau; \lambda_{k'}, \mu) \right.$$

$$\left. + (\tau + 1)^{-1} \Phi(x - \beta_k(t - \tau), \tau) \right] d\tau$$

$$+ \int_{t-1}^t \int_{-\infty}^{\infty} O(1)(t - \tau)^{-\frac{1}{2}-\frac{l}{2}}(t + 1 - \tau)^{-\frac{1}{2}} \sum_{k=1}^{s} e^{-\frac{(x-y-\lambda_k(t-\tau))^2}{C(t-\tau)}} (\delta^* + M(\tau))^2$$

$$\cdot \left[\sum_{k'=1}^{s} \theta_3(y, \tau; \lambda_{k'}, \mu) + (\tau + 1)^{-1} \Phi(y, \tau) \right] dy d\tau.$$

The first term of I_2 is estimated in the same way as in the proof of Lemma 8.7, see I_2 and I_3 in (8.40). The second term of I_2 can be written as

$$O(1)(\delta^* + M(t))^2 \sum_{k=1}^{m'} \int_0^t e^{-(t-\tau)/C}(\tau + 1)^{-\frac{3}{4}} \Phi(x - \beta_k(t - \tau), \tau) \, d\tau$$

$$= O(1)(\delta^* + M(t))^2 (t + 1)^{-\frac{3}{4}} \sum_{k=1}^{m'} \int_0^t e^{-(t-\tau)/C} \Phi(x - \beta_k(t - \tau), \tau) \, d\tau,$$

which is bounded by the right-hand side of (8.42), using Lemmas 3.9 and 3.10. Clearly I_3 is estimated in the same way.

For $2 \le l \le 5$, we have by (8.12)

$$I = \int_0^{t/2} \int_{-\infty}^{\infty} \frac{\partial^{l+1}}{\partial x^{l+1}} (G - G^*)_{ij}(x - y, t - \tau) g(y, \tau) \, dy d\tau$$

$$+ \int_{t/2}^t \int_{-\infty}^{\infty} \frac{\partial}{\partial x} (G - G^*)_{ij}(x - y, t - \tau) \frac{\partial^l}{\partial y^l} g(y, \tau) \, dy d\tau$$

$$+ \int_0^{t-1} \int_{-\infty}^{\infty} \frac{\partial^l}{\partial x^l} (G - G^*)_{ij}(x - y, t - \tau) h_y(y, \tau) \, dy d\tau$$

$$+ \int_{t-1}^t \int_{-\infty}^{\infty} \frac{\partial}{\partial x} (G - G^*)_{ij}(x - y, t - \tau) \frac{\partial^l}{\partial y^l} h(y, \tau) \, dy d\tau$$

$$= \int_0^{t/2} O(1)(t - \tau)^{-\frac{l+3}{2}} \|g(\cdot, \tau)\|_{L^1} \, d\tau + \int_0^{t/2} O(1) e^{-(t-\tau)/C} \|g(\cdot, \tau)\|_{W^{l+1,\infty}} \, d\tau$$

$$+ \int_{t/2}^{t} O(1)(t-\tau)^{-\frac{1}{2}}(t+1-\tau)^{-\frac{1}{2}} \left\| \frac{\partial^l}{\partial y^l} g(\cdot,\tau) \right\|_{L^\infty} d\tau$$

$$+ \int_{t/2}^{t} O(1)e^{-(t-\tau)/C} \left\| \frac{\partial^l}{\partial y^l} g(\cdot,\tau) \right\|_{W^{1,\infty}} d\tau$$

$$+ \int_{0}^{t-1} O(1)(t-\tau)^{-\frac{l}{2}}(t+1-\tau)^{-\frac{1}{2}} \| h_y(\cdot,\tau) \|_{L^\infty} d\tau$$

$$+ \int_{0}^{t-1} O(1)e^{-(t-\tau)/C} \| h_y(\cdot,\tau) \|_{W^{l,\infty}} d\tau$$

$$+ \int_{t-1}^{t} O(1)(t-\tau)^{-\frac{1}{2}}(t+1-\tau)^{-\frac{1}{2}} \left\| \frac{\partial^l}{\partial y^l} h(\cdot,\tau) \right\|_{L^\infty} d\tau$$

$$+ \int_{t-1}^{t} O(1)e^{-(t-\tau)/C} \left\| \frac{\partial^l}{\partial y^l} h(\cdot,\tau) \right\|_{W^{1,\infty}} d\tau.$$

By (8.25), (8.8), (8.27), (8.41), (8.38), we have

$$\| g(\cdot,\tau) \|_{L^1} = O(1)(\delta^* + M(\tau))^2 (\tau+1)^{-\frac{1}{2}},$$

$$\left\| \frac{\partial^l}{\partial y^l} g(\cdot,\tau) \right\|_{L^\infty} = O(1)(\delta^* + M(\tau))^2 \begin{cases} (\tau+1)^{-\frac{3}{2}}, & l=2,3 \\ (\tau+1)^{-1}, & l=4,5 \, , \\ (\tau+1)^{-\frac{1}{2}}, & l=6 \end{cases}$$

$$\left\| \frac{\partial^l}{\partial y^l} h(\cdot,\tau) \right\|_{L^\infty} = O(1)(\delta^* + M(\tau))^2 \begin{cases} (\tau+1)^{-\frac{3}{2}}, & l=1 \\ (\tau+1)^{-1}, & l=2,3,4 \, . \\ (\tau+1)^{-\frac{1}{2}}, & l=5,6 \end{cases}$$

Thus I is bounded by the right-hand side of (8.42). $\qquad\square$

From (8.11) and (8.14), applying Lemmas 8.1–8.8, we conclude that

$$M(t) \le C\delta^* + C(\delta^* + M(t))^2.$$

If $M(t)$ and δ^* are small, we have $M(t) \le C\delta^*$. By continuity, $M(t) \le C\delta^*$ if δ^* is small. Together with (8.15), (2.25) follows.

We have proved Theorem 2.3. From Theorem 6.15 and Remark 6.4, we can find explicitly Q_{k0} in (8.12). It is easy to see that if the left eigenspace of B associated with the zero eigenvalue is independent of u, then $Q_{k0} LB(u) = 0$ for all u, where L is the matrix consisting of the l_i. Thus when we perform a priori estimate for $\partial^l v_{ij}/\partial x^l$ using (8.11), the highest derivative of v involved in the source terms is $\partial^{l+1} v/\partial x^{l+1}$ instead of $\partial^{l+2} v/\partial x^{l+2}$. To obtain the pointwise estimate for $v(x,t)$, it is sufficient to have an L^∞ bound for $v_x(x,t)$

rather than a pointwise estimate. In fact, it is sufficient to require that $\|v_x(\cdot,t)\|_{L^\infty}$ decays as $(t+1)^{-1}$, $\|v_{xx}(\cdot,t)\|_{L^\infty}$ as $(t+1)^{-\frac{1}{2}}$, and $\|v_{xxx}(\cdot,t)\|_{L^\infty}$ is bounded. Therefore, in Theorem 2.3 it is sufficient to require $u_0 \in H^4(\mathbb{R})$ rather than $u_0 \in H^8(\mathbb{R})$. And this is Remark 2.5.

We now outline the proof of Theorem 2.6. On the right-hand side of (8.11), the first two terms give the contribution of the initial data. In (2.27) we have improved the decay rate of u_0 and u_0' to $|x|^{-\frac{3}{2}}$ as $x \to \infty$. These two terms then become

$$(8.43) \qquad O(1)\delta^*(t+1)^{-\frac{l}{2}} \sum_{k=1}^{s} \tilde{\psi}(x,t;\lambda_k)$$

for $l = 0,\ 1$, where $\tilde{\psi}$ is defined in (2.19), c.f. (4.14), (4.22). Define

$$(8.44) \quad M(t) = \sup_{0 \le \tau \le t} \max_{1 \le i \le s} \left\{ \|v_i(\cdot,\tau)\Phi_i(\cdot,\tau)^{-1}\|_{L^\infty} + \|v_{ix}(\cdot,\tau)(\tau+1)^{\frac{1}{2}}\Phi_i(\cdot,\tau)^{-1}\|_{L^\infty} \right.$$
$$+ \left\| \frac{\partial^2}{\partial x^2} v_i(\cdot,\tau) \right\|_{L^\infty} (\tau+1)^{\frac{3}{2}}$$
$$\left. + \sum_{l=2}^{3} \left(\left\| \frac{\partial^{2l-1}}{\partial x^{2l-1}} v_i(\cdot,\tau) \right\|_{L^\infty} + \left\| \frac{\partial^{2l}}{\partial x^{2l}} v_i(\cdot,\tau) \right\|_{L^\infty} \right)(\tau+1)^{\frac{4-l}{2}} \right\},$$

where

$$(8.45) \qquad \Phi_i(x,t) = \psi_{3/2}(x,t;\lambda_i) + \sum_{j \ne i} \tilde{\psi}(x,t;\lambda_j).$$

Similar to Lemma 8.8, we can show that the fourth term on the right-hand side of (8.11) is

$$(8.46) \qquad O(1)(\delta^* + M(t))^2(t+1)^{-\frac{l}{2}}\Phi_i(x,t), \quad l = 0, 1.$$

The third term has the same bound (8.46). The proof is similar to the proofs of Lemmas 8.5 and 8.7 except that we need to apply Lemma 3.4 to

$$(8.47) \quad \int_0^t \int_{-\infty}^{\infty} \frac{1}{\sqrt{4\pi\mu_{ij}(t-\tau)}} e^{-\frac{(x-y-\lambda_i(t-\tau))^2}{4\mu_{ij}(t-\tau)}}$$
$$\cdot \frac{\partial^{l+1}}{\partial y^{l+1}} \left[-l_{ij} \sum_{k \ne i} f''(0)(r_k\theta_k, r_k v_k)(y,\tau) \right] dy d\tau, \qquad l = 0, 1,$$

in (8.29), and to I_1 in (8.40). To estimate (8.47), we require $\|v_{ixx}(\cdot,t)\|_{L^\infty} = O(1)M(t)(t+1)^{-\frac{3}{2}}$. In turn we need $u_0 \in H^9(\mathbb{R})$ to close the a priori estimate.

We see that in (8.30) the contribution of the nonlinear source term

$$\frac{\partial^{l+1}}{\partial y^{l+1}} \left[-\frac{1}{2} l_{ij} \sum_{k \neq i} f''(0)(r_k \theta_k, r_k \theta_k)(y, \tau) \right], \qquad l = 0, 1,$$

gives the rates $(t+1)^{-3/2}$ and $(t+1)^{-2}$ for v_i and v_{ix}, respectively, away from the characteristic directions. These two rates can not be improved.

We now outline the proof of Theorem 2.9 when the linearization of (2.1) is not diagonalizable. By the assumption B is a nonsingular constant matrix. Thus in (8.11) $h = 0$, and in (8.12) the δ-functions disappear. For $l = 0$, 1, the first and third terms in (8.11) were estimated in Section 4, see (4.13)–(4.16), (4.31) and (4.35). The second term is estimated in a similar way as the first term, making use of (8.12), and the bounds are

$$O(1)\delta^* \sum_{k=1}^{s} \tilde{\psi}(x, t; \lambda_k) \qquad \text{and} \qquad O(1)\delta^*(t+1)^{-\frac{3}{2}},$$

respectively, for $l = 0$, 1. The fourth term is estimated in the proof of Lemma 8.8, with $h = 0$, $b = 0$, and the bounds are

$$O(1)(\delta^* + M(t))^2 \sum_{k=1}^{s} \tilde{\psi}(x, t; \lambda_k) \qquad \text{and} \qquad O(1)(\delta^* + M(t))^2(t+1)^{-\frac{3}{2}},$$

respectively, for $l = 0$, 1. Here we notice that we only need the L^∞ norm of the v_{jx} instead of the pointwise estimate, since B is a constant matrix, and that the v_{jxx} do not appear since we no longer have δ-functions. The property (2.32) then follows from (4.31) and (4.35).

At last, we point out that the solution u to (2.1), (2.2) is approximated by the solution u^* to

$$(8.48) \qquad \begin{cases} u_t + f(u)_x = B^* u_{xx}, & -\infty < x < \infty, \quad t > 0, \\ u(x, 0) = u_0(x), & -\infty < x < \infty \end{cases}$$

at a higher rate than by the diffusion waves, where

$$(8.49) \qquad B^* = (r_1, \ldots, r_s) \begin{pmatrix} l_1 B(0) r_1 & & \\ & \ddots & \\ & & l_s B(0) r_s \end{pmatrix} \begin{pmatrix} l_1 \\ \vdots \\ l_s \end{pmatrix}.$$

In fact, let

(8.50) $$u^*(x,t) = \sum_{i=1}^{s} r_i u_i^*(x,t), \qquad v_i^*(x,t) = u_i(x,t) - u_i^*(x,t).$$

Then similar to (8.11) we have

(8.51)
$$v_{ij}^*(x,t) = \int_{-\infty}^{\infty} (G - G^*)_{ij}(x - y, t)\breve{u}(y,0)\, dy$$
$$+ \int_0^t \int_{-\infty}^{\infty} \frac{1}{\sqrt{4\pi\mu_{ij}(t-\tau)}} e^{-\frac{(x-y-\lambda_i(t-\tau))^2}{4\mu_{ij}(t-\tau)}} \frac{\partial}{\partial y}(g_{ij}(u) - g_{ij}^*(u^*) + h_{ij}(u))(y,\tau)\, dy d\tau$$
$$+ \int_0^t \int_{-\infty}^{\infty} (G - G^*)_{ij}(x-y, t-\tau)\frac{\partial}{\partial y}(g(u) + h(u))(y,\tau)\, dy d\tau,$$
$$j = 1,\ldots,m_i, \quad i = 1,\ldots,s,$$

where g and h are given by (8.3). Except the contribution from $g_{ij}(u) - g_{ij}(u^*)$ in the second term, the right-hand side of (8.51) has been estimated in Lemmas 8.3, 8.4, 8.7 and 8.8; it decays faster than $v_{ij}(x,t)$ by a factor $(t+1)^{-\frac14+\varepsilon}$, where $\varepsilon > 0$ is arbitrarily small. Then by a priori estimate we have the final remark of this section:

Remark 8.9. Under the assumptions of Theorem 2.3, the solution u to (2.1), (2.2) is approximated by the solution u^* to (8.48), with

(8.52) $\quad u_i(x,t) - u_i^*(x,t) \equiv l_i u(x,t) - l_i u^*(x,t)$
$$= O(1)\delta^*\big[(t+1)^{-\frac14+\varepsilon}\Phi(x,t) + (t+1)^{-\frac12}|x - \lambda_i(t+1)|^{-1+\varepsilon}$$
$$\cdot \mathrm{char}\{(t+1)^{\frac12}/\varepsilon \le |x - \lambda_i(t+1)| \le \varepsilon(t+1)\}\big],$$

where $\varepsilon > 0$ is arbitrarily small. From (8.52) we see that $u(x,t) - u^*(x,t)$ decays as $(t+1)^{-1+\varepsilon}$ along the characteristic directions, and as $(t+1)^{-3/2+\varepsilon}$ away from those directions.

9. Applications to Continuum Mechanics

We now apply our general result to two important examples: the Navier-Stokes equations and the equations of magnetohydrodynamics.

Example 9.1. The Navier-Stokes equations describe the classical 1-D compressible, viscous flow or 1-D thermoelastic flow. In Lagrangian coordinates the equations read:

$$(9.1) \qquad \begin{cases} v_t - u_x = 0, \\ u_t + p_x = (\frac{\mu}{v} u_x)_x, \\ (e + \frac{1}{2} u^2)_t + (pu)_x = (\frac{\kappa}{v} \theta_x + \frac{\mu}{v} u u_x)_x, \end{cases}$$

where v, u, p, e and θ are, respectively, the specific volume, velocity, pressure, internal energy, and temperature of the fluid. For simplicity we assume that the viscosity μ is a nonnegative constant and the heat conductivity κ is a positive constant. Besides, we have the equation of state. Denote the entropy by s. Then among the thermodynamic variables v, p, e, θ and s, we can choose two of them as independent variables and regard the rest as smooth functions of them. We write

$$(9.2) \qquad \begin{aligned} p &= p(v,e) = \bar{p}(v,s) = \tilde{p}(v,\theta), & \theta &= \theta(v,e) = \bar{\theta}(v,s), \\ s &= s(v,e) = \tilde{s}(v,\theta), & e &= \bar{e}(v,s) = \tilde{e}(v,\theta), \end{aligned}$$

and assume

$$(9.3) \qquad \tilde{p}_v = \frac{\partial}{\partial v} \tilde{p}(v,\theta) < 0, \qquad \theta_e = \frac{\partial}{\partial e} \theta(v,e) > 0, \qquad p_e = \frac{\partial}{\partial e} p(v,e) \neq 0.$$

The thermodynamic law $de = \theta ds - p dv$ implies

$$(9.4) \qquad \begin{aligned} \bar{e}_v &= -p, & \bar{e}_s &= \theta, & \tilde{e}_\theta &= \theta \tilde{s}_\theta, \\ \tilde{e}_v &= \theta \tilde{p}_\theta - p, & s_e &= 1/\theta, & s_v &= p/\theta. \end{aligned}$$

From (9.2) and (9.4) we have

$$(9.5) \qquad \begin{aligned} \bar{p}_v &= p_v - pp_e, & \bar{\theta}_v &= -\bar{p}_s = -\frac{\tilde{p}_\theta}{\tilde{s}_\theta} = -\theta \frac{\tilde{p}_\theta}{\tilde{e}_\theta}, \\ \bar{p}_v &= \tilde{p}_v + \tilde{p}_\theta \bar{\theta}_v = \tilde{p}_v - \frac{\theta \tilde{p}_\theta^2}{\tilde{e}_\theta}, & \tilde{p}_v &= p_v + \tilde{e}_v p_e = p_v - \frac{\theta_v p_e}{\theta_e}, \\ p_e &= \bar{p}_s s_e = -\frac{\bar{\theta}_v}{\theta} = -\frac{1}{\theta}(\theta_v - p\theta_e), & \theta_v &= \bar{\theta}_v + p\theta_e = -\tilde{e}_v \theta_e. \end{aligned}$$

111

To examine that Assumptions 2.1–2.3 are satisfied. we rewrite (9.1) in the form

$$(9.6) \qquad w_t + f(w)_x = (B(w)w_x)_x.$$

Here

$$w = (v, u, e + \frac{1}{2}u^2)^t, \qquad f(w) = (-u, p, pu)^t,$$

$$(9.7) \qquad B(w) = \begin{pmatrix} 0 & 0 & 0 \\ 0 & \frac{\mu}{v} & 0 \\ \frac{\kappa}{v}\theta_v & \frac{\mu}{v}u - \frac{\kappa}{v}u\theta_e & \frac{\kappa}{v}\theta_e \end{pmatrix},$$

and we have

$$(9.8) \qquad A(w) \equiv f'(w) = \begin{pmatrix} 0 & -1 & 0 \\ p_v & -up_e & p_e \\ up_v & p - u^2 p_e & up_e \end{pmatrix}.$$

We consider w around a constant state $w^* = (v^*, 0, e^*)$ with $v^*, e^* > 0$. Thus u, $f'(0)$ and $B(0)$ is Section 2 correspond to $w - w^*$, $f(w^*)$ and $B(w^*)$ here. For Assumption 2.1, we take $U(w) = -s$ and $F(w) = 0$ as the entropy pair. Thus by (9.4),

$$\nabla U = (-s_v, us_e, -s_e) = \frac{1}{\theta}(-p, u, -1),$$

$$(9.9) \qquad A_0(w) = \nabla^2 U = \begin{pmatrix} -s_{vv} & us_{ve} & -s_{ve} \\ us_{ve} & s_e - u^2 s_{ee} & us_{ee} \\ -s_{ve} & us_{ee} & -s_{ee} \end{pmatrix}.$$

Together with (9.5) and (9.3),

$$\det A_0 = s_e(s_{vv}s_{ee} - s_{ve}^2) = \frac{1}{\theta^4}\left[-p_v\theta_e + (p\theta_e - \theta_v)\frac{\theta_v}{\theta}\right]$$

$$= \frac{1}{\theta^4}(-p_v\theta_e + p_e\theta_v) = -\frac{1}{\theta^4}\theta_e\tilde{p}_v > 0,$$

$$(9.10) \qquad -s_{vv} = -\frac{1}{s_{ee}}\left(\frac{1}{s_e}\det A_0 + s_{ve}^2\right) = \frac{1}{\theta_e}\theta^2(\theta \det A_0 + s_{ve}^2) > 0,$$

$$\det\begin{pmatrix} -s_{vv} & us_{ve} \\ us_{ve} & s_e - u^2 s_{ee} \end{pmatrix} = -s_{vv}s_e + \frac{u^2}{s_e}\det A_0 > 0.$$

Therefore A_0 is positive definite and U is strictly convex with respect to w. It is clear that $(\nabla U)f' = 0$ by (9.8) and (9.9). From (9.7) and (9.9),

$$A_0 B = \frac{1}{v}\begin{pmatrix} -\kappa\theta_v s_{ve} & \kappa u\theta_e s_{ve} & -\kappa\theta_e s_{ve} \\ \kappa u\theta_v s_{ee} & \mu s_e - \kappa u^2\theta_e s_{ee} & \kappa u\theta_e s_{ee} \\ -\kappa\theta_v s_{ee} & \kappa u\theta_e s_{ee} & -\kappa\theta_e s_{ee} \end{pmatrix},$$

which is symmetric since

$$(9.11) \qquad\qquad \theta_e s_{ve} = \theta_v s_{ee} = -\theta_v \theta_e / \theta^2.$$

It is straightforward to varify that $A_0 B$ is semi-positive definite by looking at the minors. For Assumption 2.2, we let

$$(9.12) \qquad\qquad \tilde{w} = (v, u, \theta)^t, \qquad w = f_0(\tilde{w}).$$

Hence

$$f_0'(\tilde{w}) = \begin{pmatrix} 1 & 0 & 0 \\ 0 & 1 & 0 \\ \tilde{e}_v & u & \tilde{e}_\theta \end{pmatrix}, \qquad \tilde{B} = \begin{pmatrix} 0 & 0 & 0 \\ 0 & \frac{\mu}{v} & 0 \\ 0 & \frac{\mu}{v}u & \frac{\kappa}{v} \end{pmatrix},$$

where we have used (9.5). Clearly the null space $\mathcal{N}$ of $\tilde{B}$ is independent of w or $\tilde{w}$. (However, the null space of B depends on w.) We see that $\mathcal{N}^\perp$ is spanned by $\{\eta_3\}$ or $\{\eta_2, \eta_3\}$, depending on $\mu = 0$ or $\mu > 0$, where

$$\eta_2 = (0, 1, 0)^t, \qquad \eta_3 = (0, 0, 1)^t.$$

Thus $\mathcal{N}^\perp$ is invariant under

$$f_0'(\tilde{w})^t A_0(w) = \begin{pmatrix} -s_{vv} - s_{ve}\tilde{e}_v & 0 & 0 \\ 0 & \frac{1}{\theta} & 0 \\ -s_{ve}\tilde{e}_\theta & u s_{ee}\tilde{e}_\theta & -s_{ee}\tilde{e}_\theta \end{pmatrix}.$$

It is clear that $\tilde{B}$ maps $\mathbb{R}^n$ into $\mathcal{N}^\perp$. Assumption 2.3 has already been varified in Section 6.2. We notice that the left eigenspace of $B(w)$ associated with the zero eigenvalue is independent of w. By Theorem 2.3 and Remark 2.5, if the initial data $w(x, 0) - w^* \in H^4(\mathbb{R})$ and satisfies the smallness assumptions in Theorem 2.3, then the solution to the Navier-Stokes equations has a time asymptotic state as three diffusion waves along the directions $dx/dt = \lambda_i$, $1 \le i \le 3$, where

$$\lambda_1 = -\sqrt{-\bar{p}_v^*}, \qquad \lambda_2 = 0, \qquad \lambda_3 = \sqrt{-\bar{p}_v^*}$$

are the eigenvalues of $A(w^*)$, $\bar{p}_v^*$ denotes the value of $\bar{p}_v$ at the constant state. The first and the third diffusion waves are nonlinear, corresponding to genuinely nonlinear fields of the inviscid system, while the second one is linear, corresponding to a linearly degenerate field. The difference between the solution and the diffusion waves is then given by (2.25).

Example 9.2. [LLP] In Lagrangian coordinates, the equations of magnetohydrodynamics characterizing the flow of a conducting fluid in the presence of magnetic field are

(9.13)
$$\begin{cases} v_t - u_{1x} = 0, \\ u_{1t} + [p + \frac{1}{2\mu_0}(B_2^2 + B_3^2)]_x = (\frac{\nu}{v}u_{1x})_x, \\ u_{2t} - (\frac{1}{\mu_0}B_1^* B_2)_x = (\frac{\mu}{v}u_{2x})_x, \\ u_{3t} - (\frac{1}{\mu_0}B_1^* B_3)_x = (\frac{\mu}{v}u_{3x})_x, \\ (vB_2)_t - (B_1^* u_2)_x = (\frac{1}{\sigma\mu_0 v}B_{2x})_x, \\ (vB_3)_t - (B_1^* u_3)_x = (\frac{1}{\sigma\mu_0 v}B_{3x})_x, \\ (e + \frac{1}{2}(u_1^2 + u_2^2 + u_3^2) + \frac{1}{2\mu_0}v(B_2^2 + B_3^2))_t \\ \qquad + [(p + \frac{1}{2\mu_0}(B_2^2 + B_3^2))u_1 - \frac{1}{\mu_0}B_1^*(B_2 u_2 + B_3 u_3)]_x \\ \qquad = [\frac{\nu}{v}u_1 u_{1x} + \frac{\mu}{v}(u_2 u_{2x} + u_3 u_{3x}) + \frac{\kappa}{v}\theta_x + \frac{1}{\sigma\mu_0^2 v}(B_2 B_{2x} + B_3 B_{3x})]_x, \end{cases}$$

where v, $u = (u_1, u_2, u_3)$, p, $B = (B_1^*, B_2, B_3)$, e and θ represent the specific volume, velocity, pressure, magnetic induction, internal energy and temperature respectively, $p = p(v, e)$, $\theta = \theta(v, e)$, B_1^* is a constant, μ, $\nu \geq 0$ are the two coefficients of viscosity, $\mu_0 > 0$ is the magnetic permeability, $\kappa > 0$ the heat conductivity and $1/\sigma \geq 0$ the electrical resistivity. Again, we have the equation of state.

Denote the entropy by s. Introduce the notations in (9.2) and assume (9.3). We have the thermodynamics relations in (9.4) and (9.5). Denote the total energy and full pressure, respectively, as

(9.14)
$$E = e + \frac{1}{2}(u_1^2 + u_2^2 + u_3^2) + \frac{v}{2\mu_0}(B_2^2 + B_3^2),$$
$$P = p + \frac{1}{2\mu_0}(B_2^2 + B_3^2).$$

We write (9.13) in the form of (9.6),

$$w_t + f(w)_x = (B(w)w_x)_x,$$

with

$$w = (v, u_1, u_2, u_3, vB_2, vB_3, E)^t,$$
$$f(w) = (-u_1, P, -\frac{1}{\mu_0}B_1^* B_2, -\frac{1}{\mu_0}B_1^* B_3, -B_1^* u_2, -B_1^* u_3, Pu_1 - \frac{1}{\mu_0}B_1^*(B_2 u_2 + B_3 u_3))^t,$$

$$(9.15a) \qquad B(w) = \begin{pmatrix} 0 & 0 & 0 & 0 & 0 & 0 & 0 \\ 0 & \frac{\nu}{v} & 0 & 0 & 0 & 0 & 0 \\ 0 & 0 & \frac{\mu}{v} & 0 & 0 & 0 & 0 \\ 0 & 0 & 0 & \frac{\mu}{v} & 0 & 0 & 0 \\ -\frac{B_2}{\sigma\mu_0 v^2} & 0 & 0 & 0 & \frac{1}{\sigma\mu_0 v^2} & 0 & 0 \\ -\frac{B_3}{\sigma\mu_0 v^2} & 0 & 0 & 0 & 0 & \frac{1}{\sigma\mu_0 v^2} & 0 \\ a_1 & a_2 u_1 & a_3 u_2 & a_3 u_3 & a_4 B_2 & a_4 B_3 & b \end{pmatrix},$$

where

$$(9.15b) \qquad \begin{aligned} & b = \frac{\kappa\theta_e}{v}, \qquad a_1 = \frac{\kappa\theta_v}{v} + (B_2^2 + B_3^2)\left(\frac{b}{2\mu_0} - \frac{1}{\sigma\mu_0^2 v^2}\right), \\ & a_2 = \frac{\nu}{v} - b, \qquad a_3 = \frac{\mu}{v} - b, \qquad a_4 = -\frac{b}{\mu_0} + \frac{1}{\sigma\mu_0^2 v^2}. \end{aligned}$$

Clearly,

$$(9.16a) \qquad A(w) \equiv f'(w) = \begin{pmatrix} 0 & -1 & 0 & 0 & 0 & 0 & 0 \\ c_1 & -p_e u_1 & -p_e u_2 & -p_e u_3 & c_2 B_2 & c_2 B_3 & p_e \\ c_3 B_2 & 0 & 0 & 0 & -c_3 & 0 & 0 \\ c_3 B_3 & 0 & 0 & 0 & 0 & -c_3 & 0 \\ 0 & 0 & -B_1^* & 0 & 0 & 0 & 0 \\ 0 & 0 & 0 & -B_1^* & 0 & 0 & 0 \\ c_4 & c_5 & c_6 & c_7 & c_8 & c_9 & p_e u_1 \end{pmatrix},$$

where

$$(9.16b) \qquad \begin{aligned} & c_1 = p_v + \frac{1}{\mu_0}(B_2^2 + B_3^2)\left(\frac{p_e}{2} - \frac{1}{v}\right), \qquad c_2 = \frac{1}{\mu_0}\left(\frac{1}{v} - p_e\right), \qquad c_3 = \frac{B_1^*}{\mu_0 v}, \\ & c_4 = c_1 u_1 + \frac{B_1^*}{\mu_0 v}(B_2 u_2 + B_3 u_3), \qquad c_5 = P - p_e u_1^2, \qquad c_6 = -p_e u_1 u_2 - \frac{B_1^* B_2}{\mu_0}, \\ & c_7 = -p_e u_1 u_3 - \frac{B_1^* B_3}{\mu_0}, \qquad c_8 = c_2 u_1 B_2 - \frac{B_1^* u_2}{\mu_0 v}, \qquad c_9 = c_2 u_1 B_3 - \frac{B_1^* u_3}{\mu_0 v}. \end{aligned}$$

We consider w around a constant state

$$(9.17) \qquad w^* = (v^*, 0, 0, 0, v^* B_2^*, v^* B_3^*, E^*)^t,$$

where $E^* = e^* + (B_2^{*2} + B_3^{*2})v^*/(2\mu_0)$, $v^* > 0$, $e^* > 0$, B_2^* and B_3^* are constants.

In Assumption 2.1 again we take $U(w) = -s$ and $F(w) = 0$ as the entropy pair. Therefore,

$$(9.18a) \qquad \nabla U = -s_v \varepsilon_1 - s_e \nabla e,$$

$$(9.18b) \qquad A_0 \equiv \nabla^2 U = -s_{vv} \eta_1^t \eta_1 - s_{ee}(\nabla e)^t \nabla e - s_e \nabla^2 e - s_{ve}((\nabla e)^t \eta_1 + \eta_1^t \nabla e),$$

where

$$\eta_1 = (1, 0, \ldots, 0),$$

$$\nabla e = \left(\frac{1}{2\mu_0} (B_2^2 + B_3^2), -u_1, -u_2, -u_3, -\frac{1}{\mu_0} B_2, -\frac{1}{\mu_0} B_3, 1 \right),$$

(9.19)
$$\nabla^2 e = \begin{pmatrix} -\frac{B_2^2+B_3^2}{\mu_0 v} & 0 & 0 & 0 & \frac{B_2}{\mu_0 v} & \frac{B_3}{\mu_0 v} & 0 \\ 0 & -1 & 0 & 0 & 0 & 0 & 0 \\ 0 & 0 & -1 & 0 & 0 & 0 & 0 \\ 0 & 0 & 0 & -1 & 0 & 0 & 0 \\ \frac{B_2}{\mu_0 v} & 0 & 0 & 0 & -\frac{1}{\mu_0 v} & 0 & 0 \\ \frac{B_3}{\mu_0 v} & 0 & 0 & 0 & 0 & -\frac{1}{\mu_0 v} & 0 \\ 0 & 0 & 0 & 0 & 0 & 0 & 0 \end{pmatrix}.$$

Clearly the sum of the first three terms on the right-hand side of (9.18b) is positive definite by (9.10), (9.4) and (9.3). Notice the special form of the last term. To show (9.18b) is positive definite, we only need to show $\det A_0 > 0$. By eliminating the first to sixth rows of the second term on the right, we have

$$\det A_0 = \frac{s_e^5}{\mu_0^2 v^2} (s_{vv} s_{ee} - s_{ve}^2) > 0,$$

c.f. (9.10). Hence U is strictly convex. From (9.16), (9.18a), (9.19) and (9.4) we can varify $(\nabla U) f' = 0$. From (9.19) and (9.15) we have by direct computation that

$$(\nabla e) B = b \nabla e + \frac{\kappa}{v} \theta_v \eta_1.$$

Thus (9.18b), (9.11) and (9.15b) imply that

(9.20)
$$\begin{aligned} A_0 B &= [-s_{ee}(\nabla e)^t - s_{ve} \eta_1^t](\nabla e) B - s_e (\nabla^2 e) B \\ &= -\frac{\kappa}{v} s_{ee} \theta_e (\nabla e + \frac{\theta_v}{\theta_e} \eta_1)^t (\nabla e + \frac{\theta_v}{\theta_e} \eta_1) - s_e (\nabla^2 e) B. \end{aligned}$$

Clearly the first term on the right is symmetric semi-positive definite. By direct computation, the second term is

$$-s_e \begin{pmatrix} -\frac{B_2^2+B_3^2}{\sigma \mu_0^2 v^3} & 0 & 0 & 0 & \frac{B_2}{\sigma \mu_0^2 v^3} & \frac{B_3}{\sigma \mu_0^2 v^3} & 0 \\ 0 & -\frac{\nu}{v} & 0 & 0 & 0 & 0 & 0 \\ 0 & 0 & -\frac{\mu}{v} & 0 & 0 & 0 & 0 \\ 0 & 0 & 0 & -\frac{\mu}{v} & 0 & 0 & 0 \\ \frac{B_2}{\sigma \mu_0^2 v^3} & 0 & 0 & 0 & -\frac{1}{\sigma \mu_0^2 v^3} & 0 & 0 \\ \frac{B_3}{\sigma \mu_0^2 v^3} & 0 & 0 & 0 & 0 & -\frac{1}{\sigma \mu_0^2 v^3} & 0 \\ 0 & 0 & 0 & 0 & 0 & 0 & 0 \end{pmatrix},$$

which is also symmetric semi-positive definite. Hence $A_0 B$ is symmetric semi-positive definite.

For Assumption 2.2 we let

(9.21)
$$\tilde{w} = (v, u_1, u_2, u_3, B_2, B_3, \theta)^t, \qquad w = f_0(\tilde{w}).$$

Then

(9.22)
$$f_0'(\tilde{w}) = \begin{pmatrix} 1 & 0 & 0 & 0 & 0 & 0 & 0 \\ 0 & 1 & 0 & 0 & 0 & 0 & 0 \\ 0 & 0 & 1 & 0 & 0 & 0 & 0 \\ 0 & 0 & 0 & 1 & 0 & 0 & 0 \\ B_2 & 0 & 0 & 0 & v & 0 & 0 \\ B_3 & 0 & 0 & 0 & 0 & v & 0 \\ \frac{B_2^2+B_3^2}{2\mu_0} + \tilde{e}_v & u_1 & u_2 & u_3 & \frac{B_2 v}{\mu_0} & \frac{B_3 v}{\mu_0} & \tilde{e}_\theta \end{pmatrix},$$

(9.23)
$$\tilde{B}(\tilde{w}) \equiv B(w)f_0'(\tilde{w}) = \begin{pmatrix} 0 & 0 & 0 & 0 & 0 & 0 & 0 \\ 0 & \frac{\nu}{v} & 0 & 0 & 0 & 0 & 0 \\ 0 & 0 & \frac{\mu}{v} & 0 & 0 & 0 & 0 \\ 0 & 0 & 0 & \frac{\mu}{v} & 0 & 0 & 0 \\ 0 & 0 & 0 & 0 & \frac{1}{\sigma\mu_0 v} & 0 & 0 \\ 0 & 0 & 0 & 0 & 0 & \frac{1}{\sigma\mu_0 v} & 0 \\ 0 & \frac{\nu}{v}u_1 & \frac{\mu}{v}u_2 & \frac{\mu}{v}u_3 & \frac{B_2}{\sigma\mu_0^2 v} & \frac{B_3}{\sigma\mu_0^2 v} & \frac{\kappa}{v} \end{pmatrix}.$$

Clearly the null space $\mathcal{N}$ of $\tilde{B}(\tilde{w})$ is independent of $\tilde{w}$. From (9.22), (9.18b), (9.4), (9.5) and (9.11) we have

$$f_0'(\tilde{w})^t A_0(w) = \begin{pmatrix} -\frac{\tilde{p}_v}{\theta} & 0 & 0 & 0 & 0 & 0 & 0 \\ 0 & \frac{1}{\theta} & 0 & 0 & 0 & 0 & 0 \\ 0 & 0 & \frac{1}{\theta} & 0 & 0 & 0 & 0 \\ 0 & 0 & 0 & \frac{1}{\theta} & 0 & 0 & 0 \\ -\frac{B_2}{\mu_0\theta} & 0 & 0 & 0 & \frac{1}{\mu_0\theta} & 0 & 0 \\ -\frac{B_3}{\mu_0\theta} & 0 & 0 & 0 & 0 & \frac{1}{\mu_0\theta} & 0 \\ -\tilde{e}_\theta s_{ve} + \frac{B_2^2+B_3^2}{2\mu_0\theta^2} & -\frac{u_1}{\theta^2} & -\frac{u_2}{\theta^2} & -\frac{u_3}{\theta^2} & -\frac{B_2}{\mu_0\theta^2} & -\frac{B_3}{\mu_0\theta^2} & \frac{1}{\theta^2} \end{pmatrix}.$$

Notice that we have assumed $\kappa > 0$. Hence $\mathcal{N}^\perp$ is invariant under $f_0'(\tilde{w})^t A_0(w)$. It is also clear that the range of $\tilde{B}(\tilde{w})$ is in $\mathcal{N}^\perp$.

Assumption 2.3 has been varified in Section 6.2 for the following cases:

(i) $\nu > 0$, $\mu > 0$, and $1/\sigma > 0$ (finitely conducting);

(ii) $\nu = \mu = 0$, and $1/\sigma > 0$, while $B_1^* \neq 0$;

(iii) $\nu > 0$, $\mu > 0$, and $1/\sigma = 0$ (perfectly conducting), while $B_1^* \neq 0$.

For these cases our main result applys. Again we notice that the left eigenspace of $B(w)$ associated with the zero eigenvalue is independent of w. Thus if $w(x,0) - w^* \in H^4(\mathbb{R})$ and satisfies the smallness assumptions in Theorem 2.3, the initial value problem of the equations of magnetohydrodynamics has an asymptotic solution which is a superposition of diffusion waves. From Section 6.2 $A(w^*)$ has eigenvalues in nondecreasing order as

$$(9.24) \qquad\qquad -c_f, -c_a, -c_s, 0, c_s, c_a, c_f,$$

where c_f, c_a, c_s are the fast, the Alfvén, and the slow wave speeds, respectively, given by (6.63). If $B_1^* = 0$, $c_s = c_a = 0$. In this case we have a multiple-mode diffusion wave along the particle path $dx/dt = 0$, and two single-mode diffusion waves along the fast wave directions. If $B_1^* \neq 0$ and $B_2^* = B_3^* = 0$, either $c_f = c_a$ or $c_s = c_a$. In this case we have a single-mode diffusion wave along the particle path, and two multiple-mode diffusion waves along the Alfvén wave directions. In addition we have two single-mode diffusion waves along the fast (slow) wave directions if $-\bar{p}_v > c_a^2$ $(-\bar{p}_v < c_a^2)$, where $\bar{p}_v$ takes value at the constant state. If $B_1^* \neq 0$ and $B_2^{*2} + B_3^{*2} \neq 0$, all the eigenvalues in (9.24) are distinct, and we have single-mode diffusion waves in all the characteristic directions. The difference between the solution and the asymptotic solution is then given by (2.25).

REFERENCES

[BW]   M. Brio and C. C. Wu, *An upwind differencing scheme for the equations of ideal magnetohydro-dynamics*, J. Comput. Phys. **75** (1988), 400–422.

[Bu]   J. B. Butler, Jr., *Perturbation series for eigenvalues of analytic non-symmetric operators*, Arch. Math. **10** (1959), 21–27.

[Ch]   I-L. Chern, *Multiple-mode diffusion waves for viscous nonstrictly hyperbolic conservation laws*, Comm. Math. Phys. **138** (1991), 51–61.

[Co]   J. D. Cole, *On a quasi-linear parabolic equation occurring in aerodynamics*, Quart. Appl. Math. **9** (1951), 225–236.

[Fr]   H. Freistühler, *Rotational degeneracy of hyperbolic systems of conservation laws*, Arch. Rational Mech. Anal. **113** (1991), 39–64.

[FL]   K. O. Friedrichs and P. D. Lax, *Systems of conservation equations with a convex extension*, Proc. Nat. Acad. Sci. U.S.A. **68** (1971), 1686–1688.

[H]   D. Hoff, *Construction of solutions for compressible, isentropic Navier-Stokes equations in one space dimension with nonsmooth initial data*, Proc. Roy. Soc. Edinburgh **103A** (1986), 301–315.

[Ho]   E. Hopf, *The partial differential equation $u_t + uu_x = \mu u_{xx}$*, Comm. Pure Appl. Math. **3** (1950), 201–230.

[Kt]   T. Kato, *Perturbation theory for linear operators*, 2nd edn., Springer, New York, 1976.

[K]   S. Kawashima, *Systems of a hyperbolic-parabolic composite type, with applications to the equations of magnetohydrodynamics*, Doctoral thesis, Kyoto University, 1983.

[Ka]   S. Kawashima, *Large-time behavior of solutions to hyperbolic-parabolic systems of conservation laws and applications*, Proc. Roy. Soc. Edinburgh **106A** (1987), 169–194.

[KO]   S. Kawashima and M. Okada, *Smooth global solutions for the one-dimensional equations in magnetohydrodynamics*, Proc. Japan Acad. Ser. A **58** (1982), 384–387.

[LLP] L. D. Landau, E. M. Lifshitz and L. P. Pitaevskiĭ, *Electrodynamics of continuous media*, 2nd edn., Pergamon Press, New York, 1984.

[La]   P. D. Lax, *Hyperbolic systems of conservation laws, II*, Comm. Pure Appl. Math. **10** (1957), 537-566.

[L]   T.-P. Liu, *Nonlinear stability of shock waves for viscous conservation laws*, Mem. Am. Math. Soc. **56** (1985), no. 328.

119

[Liu]	T.-P. Liu, *Interactions of nonlinear hyperbolic waves*, Nonlinear Analysis (F.-C. Liu and T.-P. Liu, eds.), World Scientific, Singapore, 1991, pp. 171–184.

[SK]	Y. Shizuta and S. Kawashima, *Systems of equations of hyperbolic-parabolic type with applications to the discrete Boltzmann equation*, Hokkaido Math. J. **14** (1985), 249–275.

[Sm]	J. Smoller, *Shock waves and reaction-diffusion equations*, Springer-Verlag, New York, 1983.

[UKS]	T. Umeda, S. Kawashima and Y. Shizuta, *On the decay of solutions to the linearized equations of electro-magneto-fluid dynamics*, Japan J. Appl. Math. **1** (1984), 435–457.

[Ze]	Y. Zeng, L^1 *asymptotic behavior of compressible, isentropic, viscous 1-D flow*, Comm. Pure Appl. Math. **47** (1994), 1053–1082.

[Z]	Y. Zeng, L^p *asymptotic behavior of solutions to hyperbolic-parabolic systems of conservation laws*, Arch. Math. (to appear).

DEPARTMENT OF MATHEMATICS, STANFORD UNIVERSITY, STANFORD, CA 94305

DIVISION OF APPLIED MATHEMATICS, BROWN UNIVERSITY, PROVIDENCE, RI 02912

Editorial Information

To be published in the *Memoirs*, a paper must be correct, new, nontrivial, and significant. Further, it must be well written and of interest to a substantial number of mathematicians. Piecemeal results, such as an inconclusive step toward an unproved major theorem or a minor variation on a known result, are in general not acceptable for publication. *Transactions* Editors shall solicit and encourage publication of worthy papers. Papers appearing in *Memoirs* are generally longer than those appearing in *Transactions* with which it shares an editorial committee.

As of September 30, 1996, the backlog for this journal was approximately 7 volumes. This estimate is the result of dividing the number of manuscripts for this journal in the Providence office that have not yet gone to the printer on the above date by the average number of monographs per volume over the previous twelve months, reduced by the number of issues published in four months (the time necessary for preparing an issue for the printer). (There are 6 volumes per year, each containing at least 4 numbers.)

A Copyright Transfer Agreement is required before a paper will be published in this journal. By submitting a paper to this journal, authors certify that the manuscript has not been submitted to nor is it under consideration for publication by another journal, conference proceedings, or similar publication.

Information for Authors and Editors

Memoirs are printed by photo-offset from camera copy fully prepared by the author. This means that the finished book will look exactly like the copy submitted.

The paper must contain a *descriptive title* and an *abstract* that summarizes the article in language suitable for workers in the general field (algebra, analysis, etc.). The *descriptive title* should be short, but informative; useless or vague phrases such as "some remarks about" or "concerning" should be avoided. The *abstract* should be at least one complete sentence, and at most 300 words. Included with the footnotes to the paper, there should be the 1991 *Mathematics Subject Classification* representing the primary and secondary subjects of the article. This may be followed by a list of *key words and phrases* describing the subject matter of the article and taken from it. A list of the numbers may be found in the annual index of *Mathematical Reviews*, published with the December issue starting in 1990, as well as from the electronic service e-MATH [**telnet e-MATH.ams.org** (or **telnet 130.44.1.100**). Login and password are **e-math**]. For journal abbreviations used in bibliographies, see the list of serials in the latest *Mathematical Reviews* annual index. When the manuscript is submitted, authors should supply the editor with electronic addresses if available. These will be printed after the postal address at the end of each article.

Electronically prepared papers. The AMS encourages submission of electronically prepared papers in $\mathcal{A}\mathcal{M}\mathcal{S}$-TEX or $\mathcal{A}\mathcal{M}\mathcal{S}$-LATEX. The Society has prepared author packages for each AMS publication. Author packages include instructions for preparing electronic papers, the *AMS Author Handbook*, samples, and a style file that generates the particular design specifications of that publication series for both $\mathcal{A}\mathcal{M}\mathcal{S}$-TEX and $\mathcal{A}\mathcal{M}\mathcal{S}$-LATEX.

Authors with FTP access may retrieve an author package from the Society's Internet node **e-MATH.ams.org** (130.44.1.100). For those without FTP access, the author package can be obtained free of charge by sending e-mail to `pub@math.ams.org` (Internet) or from the Publication Division, American Mathematical Society, P.O. Box 6248, Providence, RI 02940-6248. When requesting an author package, please specify $\mathcal{A}_{\mathcal{M}}\mathcal{S}$-TEX or $\mathcal{A}_{\mathcal{M}}\mathcal{S}$-LATEX, Macintosh or IBM (3.5) format, and the publication in which your paper will appear. Please be sure to include your complete mailing address.

Submission of electronic files. At the time of submission, the source file(s) should be sent to the Providence office (this includes any TEX source file, any graphics files, and the DVI or PostScript file).

Before sending the source file, be sure you have proofread your paper carefully. The files you send must be the EXACT files used to generate the proof copy that was accepted for publication. For all publications, authors are required to send a printed copy of their paper, which exactly matches the copy approved for publication, along with any graphics that will appear in the paper.

TEX files may be submitted by email, FTP, or on diskette. The DVI file(s) and PostScript files should be submitted only by FTP or on diskette unless they are encoded properly to submit through e-mail. (DVI files are binary and PostScript files tend to be very large.)

Files sent by electronic mail should be addressed to the Internet address `pub-submit@math.ams.org`. The subject line of the message should include the publication code to identify it as a Memoir. TEX source files, DVI files, and PostScript files can be transferred over the Internet by FTP to the Internet node `e-math.ams.org` (130.44.1.100).

Electronic graphics. Figures may be submitted to the AMS in an electronic format. The AMS recommends that graphics created electronically be saved in Encapsulated PostScript (EPS) format. This includes graphics originated via a graphics application as well as scanned photographs or other computer-generated images.

If the graphics package used does not support EPS output, the graphics file should be saved in one of the standard graphics formats—such as TIFF, PICT, GIF, etc.—rather than in an application-dependent format. Graphics files submitted in an application-dependent format are not likely to be used. No matter what method was used to produce the graphic, it is necessary to provide a paper copy to the AMS.

Authors using graphics packages for the creation of electronic art should also avoid the use of any lines thinner than 0.5 points in width. Many graphics packages allow the user to specify a "hairline" for a very thin line. Hairlines often look acceptable when proofed on a typical laser printer. However, when produced on a high-resolution laser imagesetter, hairlines become nearly invisible and will be lost entirely in the final printing process.

Screens should be set to values between 15% and 85%. Screens which fall outside of this range are too light or too dark to print correctly.

Any inquiries concerning a paper that has been accepted for publication should be sent directly to the Editorial Department, American Mathematical Society, P. O. Box 6248, Providence, RI 02940-6248.

Selected Titles in This Series

(Continued from the front of this publication)

(See the AMS catalog for earlier titles)